D0108436

REVOLUTIONS IN SCIENCE

Eureka!

The Birth of Science

Andrew Gregory

Series editor: Jon Turney

Published in the UK in 2001
by Icon Books Ltd., Grange Road,
Duxford, Cambridge CB2 4QF
E-mail: info@iconbooks.co.uk
www.iconbooks.co.uk

Sold in the UK, Europe, South Africa
and Asia by Faber and Faber Ltd.,
3 Queen Square, London WC1N 3AU
or their agents

Distributed in the UK, Europe,
South Africa and Asia by
Macmillan Distribution Ltd.,
Houndmills, Basingstoke RG21 6XS

Published in Australia in 2001
by Allen & Unwin Pty. Ltd.,
PO Box 8500, 83 Alexander Street,
Crows Nest, NSW 2065

Published in the USA in 2001
by Totem Books
Inquiries to: Icon Books Ltd.,
Grange Road, Duxford,
Cambridge CB2 4QF, UK

Distributed to the trade in the USA
by National Book Network Inc.,
4720 Boston Way, Lanham,
Maryland 20706

Distributed in Canada by
Penguin Books Canada,
10 Alcorn Avenue, Suite 300,
Toronto, Ontario M4V 3B2

ISBN 1 84046 289 2

Series editor: Jon Turney

Originating editor: Simon Flynn

Typesetting by Hands Fotoset

Printed and bound in the UK by
Mackays of Chatham plc

Contents

Acknowledgements

I would like to thank the editors, Jon Turney and Simon Flynn, for their patience and efficiency and for their comments on the manuscript. I would also like to thank Ms. Sheelagh Doherty RGN, RSCN, RM for her support, her comments and for checking the manuscript for medical accuracy. Without their help this would have been a less interesting and less accurate book.

Dedication

For Sheelagh, *with love*

The Creation of Science

Science has done much to shape both the world we live in and the way in which we think about that world. But what are the origins of science? What came before science, and how and why was that transformed into a new and progressive way of thinking about and investigating our world? Who were the people who effected this transformation? When and where did science begin?

Prior to science, there was technology. People knew how to do many useful things, without understanding quite why they happened, or why natural phenomena occurred. When they attempted to explain their world, it was in terms of myths and anthropomorphic gods. So thunder, lightning, earthquakes and disease were all due to the actions of the gods, while the origins of the world and human beings were a matter of myth. These myths often involved the sexual coupling of the gods – such as those of sea and sky to create earth – since procreation was one of the few models for the production of something new that the ancients possessed. These gods were supposed to have many human fallibilities. They sometimes acted in anger, jealousy or spite, and their actions were often unpredictable to humans.

A good example here is the daily passage of the sun across the sky. What do we see, and how do we explain it? To a pre-scientific society, the sun might well be a god driving his chariot across the heavens. Many primitive cosmologies supposed the universe to be hemispherical. There was a flat earth with a hemispherical bowl of the heavens above it. So the sun would disappear in the evening and reappear each morning, but what happened in between was a mystery – the subject of myth. Many ancient societies could accurately predict the time that the sun would rise, and at what point on the horizon. Any sort of scientific explanation of the sun or its motions, though, was beyond them.

At some point, a new and more critical attitude came about. People began to reject myths and explanations in terms of the gods as arbitrary and fanciful. Instead, they began to use theories for which they could gather evidence and debate the merits. They considered their world to be a natural place, in the sense that it was free from supernatural intervention, and so in need of *natural* explanations. Thunder and lightning were to be explained in terms of storm clouds, and not the anger of the gods. The world was now seen as a place where events happened in a regular and predictable manner, and were not dependent on the whims of the gods.

In many ways, it is remarkable that science came about at all. Science is not a 'natural' activity in the sense that it comes easily or instinctively to humans.

Technology, the ability to manipulate our environment to our benefit, may come relatively easily; but science, involving understanding and explaining our world, does not. Nor is science a 'natural' way of thinking, as we can see from the fact that the first societies were dominated by myth and anthropomorphic deities. One might also consider the prevalence of non-scientific thought in the world today. Nor is science merely applied common sense. Many of the ideas of science, even at its very outset, have been quite contrary to common sense. Nor, one must say, was science a productive activity in the sense that it would reap immediate material benefits. So why, and how, did the transition to scientific thinking occur? Who was responsible for it? These are the questions that this book will investigate.

When and where this transformation occurred is relatively easy to pin down. The first steps towards scientific explanation were taken in ancient Greece around 600 BC. Prior to that, the Babylonians and the Egyptians had evolved advanced technologies, but had not progressed beyond mythological explanations. The Greeks drew deeply on these technologies, especially in astronomy, geometry and medicine, and began to produce the first crude theories of how the world might work in an entirely natural manner. This book will follow the Greeks on their adventure in this new type of thinking, looking at the ideas and approaches that they created, and the increasing sophistication of their theories. It will also look at the social background that allowed them to initiate and develop a radically new way of looking at the world.

This book is not a comprehensive treatment of Greek science. That would require a work many times longer than this. Rather, it attempts to capture the essence and the spirit of the Greek achievement, and something of the excitement of the debate between the Greek thinkers. It attempts to convey what the Greeks thought their world was like, and how they went about investigating it. The Greek picture of the world is of great relevance for several reasons:

i) It was formative for virtually all Western thought down to the scientific revolution of the seventeenth century, not merely in science but in philosophy and religion as well. The dominant mode of thought – and most of the alternatives to it – was a combination of Greek science, Greek philosophy and Christian theology (which in turn was deeply influenced by Greek philosophy and theology). To understand the nature of the scientific revolution, one needs to understand Greek science and its strengths and weaknesses.
ii) The influence of Greek ideas did not come to an end with the scientific revolution. Many of their ideas, such as atomism, are still alive and well, and many of the principles laid down by the Greeks for under-

standing and investigating the universe are still valid today.

iii) Greek science displays fascinating differences from modern science. The spirit of investigation may have remained the same, but the content of science has changed radically. The Greeks had many wonderful ideas, but some look somewhat strange to the modern eye, and some are wrong. That is no great surprise, since we are talking of the pioneers of science, and a time gap of over two millennia. What I shall try to explain is why, given the resources available to the Greeks, intelligent people would have found these ideas attractive. Often the Greeks had very good reasons for their odder beliefs.

The project begun by the ancient Greeks is one that has deeply affected every aspect of thinking in the modern world, and every aspect of our lives. Our conception of the natural world traces its ancestry to the ancient Greeks, and our science has its roots there. This book is the story of the origins of a great quest to understand the world we live in, a quest that continues today and that still owes a great deal to its originators.

1 The Early Greeks and their Predecessors

Science must begin with myths, and with the criticism of myths.

Sir Karl Popper, 'The Philosophy of Science', in C.A. Mace (ed.), *British Philosophy in the Mid-Century* (1957)

When and where science began depends to some extent on what we think science is. Science is a more sophisticated activity than technology. With technology, one knows *how* to do something, or *when* something will occur. With science, one has a theory and an explanation of *why* such a thing should happen. A good example is the production of iron tools. One can mine iron ore and go through the processes of refining and forging iron without having any idea of the nature of those processes or why they work. If so, one has only technology. Or one might have a theory which allows one to explain each step of the process, and so understand what is going on. One might then be said to have science. Many ancient societies clearly had technology. Indeed, we define certain historical ages by the sort of technology that was possessed. So we have the Stone Age, Bronze Age, Iron Age etc., characterised by the technology for producing stone, bronze and iron tools. All societies have had some form of technology. We might even say that some animals have a rudimentary technology, since they use tools (e.g., birds using stones to break open snail shells). We would not say, though, that they have science. Science is a step beyond technology, requiring at least the *attempt* to explain and understand.

How much of a step beyond is another matter. The more one builds into a definition of science, the later one is likely to believe that it begins. As a basic minimum, though, we are looking for the following:

i) Science deals with the natural world, so we are looking for an awareness of a distinction between the natural and the supernatural, and a desire to explain using only natural factors and not, for example, the intervention of the gods.
ii) Science is expressed in terms of theories, so we are looking for theories about the world, as opposed to the myths or poems typical of some ancient societies.
iii) Science is also characterised by the use of mathematics, experiment and observation. We are looking for science as opposed to mere technology.
iv) It would also be helpful if our candidates as the originators of science were aware of the differences

between what they were doing and what their predecessors were doing.

All of this, and rather more, we find with the ancient Greeks but not with any previous society. We are not, of course, looking for something that is identical with modern science. Science progresses, and we can hardly expect the content of ancient theories to stand comparison with modern theories. What we are looking for is something which has enough in common with modern science, in terms of orientation of investigation and the types of explanation offered, to be recognisable as its ancestor.

Science Must Begin with Myths

It is impossible to doubt the technological and mathematical achievements of some societies prior to the Greeks. The Babylonians had in place a sophisticated number system and means of solving equations. A great number of clay tablets have been found in tombs in what was Mesopotamia and is now Iraq (between the Tigris and Euphrates rivers), giving us a considerable insight into the achievements of the Babylonians. They managed to construct a workable calendar, by no means a trivial feat when starting from scratch. The relation between days, lunar months and solar years is a very complex one. There is not a whole number of days in a lunar month or solar year, nor a whole number of lunar months in a solar year. The Babylonians were also very good observers of the heavens. Some of their clay tablets turned out to be detailed and accurate observations of the movements of the heavens, with astronomical predictions worked out mathematically. The best candidate for science prior to the ancient Greeks is undoubtedly Babylonian astronomy.

While the Babylonians were very good at observation and prediction, they never got beyond describing the heavens in terms of myths and poems. They had no theories as to the nature of the heavens, and they failed to produce any explanations of the phenomena. Their predictions worked by extrapolation from the data, rather than from a model of the heavens. For example, if an eclipse has happened in year 1, year 3, year 5 and year 7, then we might predict that an eclipse will also happen in year 9. One can make this prediction without saying anything at all about the nature of the heavens or the nature of eclipses. In fairness to the Babylonians, the mathematics they used was much more sophisticated than this, as were their predictions. But there was nothing which explained what an eclipse was, or why it should happen at a particular time. The Babylonians were concerned only with *when*, not *why* – they had a technology of astronomical prediction, but not the science of astronomy. They also had a purely mythical cosmogony (theory of the creation of the universe).

Here is a passage from the *Enuma Elisha*, the Babylonian creation epic, probably composed early in the second millennium BC:

> *When the upper heavens had as yet no name,*
> *And the lower heavens had not as yet been named,*
> *When only the primeval Apsu which was to beget them yet existed,*
> *And their mother Ti'amat, who gave birth to them all:*
> *When all was as yet mixed in the waters,*
> *And no dry land could be seen – not even a marsh;*
> *When none of the Gods had yet been brought into existence,*
> *Or been given names, or had their destinies fixed:*
> *Then were the Gods created between the begetters.*
>
> S. Toulmin and J. Goodfield,
> *The Fabric of the Heavens* (1961), p. 42.

With the Greeks came a new sort of society with some radically different attitudes to the world and how to explain it.

Two Cultures

Greek society was affluent enough for some relatively well-to-do people to have sufficient time to consider philosophical questions, including the nature of the world around them. They also had the intellectual freedom to pursue original lines of thought. Significantly, the Greeks had no centralised religion and no official caste of priests. Babylonian society was hierarchical, both in the literal sense that it was ruled by priests (from the Greek *hieros*, priest, *arche*, rule) and in the sense that it was rigidly stratified. In Greek society there was tolerance of a wide range of religious views, and of debate in general. An excellent example of this can be found in the views of Xenophanes of Colophon (*fl.* 520 BC),[1] a philosopher and theologian. Xenophanes lived to be at least 92, wrote in verse and travelled extensively. He was particularly critical of popular religion, and of the gods in the epic tales of Homer and Hesiod. He said that:

> *Homer and Hesiod have ascribed to the gods all those things which are shameful and reproachful among men: theft, adultery and deceiving each other . . . Mortals believe that the gods are born, and that they have clothes, speech and bodies similar to their own . . . If cattle, horses and lions had hands, and could draw with those hands and accomplish the works of men, horses would draw the forms of gods as like horses, and cattle like cattle, and each would make their bodies as each had themselves . . . The*

1. With some of the earlier Greeks, we have only a rough indication of when they did their most important work, rather than specific dates for birth and death; '*fl.*' (*floruit*, when they flourished) indicates this date.

Ethiopians claim their gods are snub-nosed and black, while the Thracians claim theirs have blue eyes and red hair.

Here I should explain that in the case of some Greek thinkers, notably Plato and Aristotle, we are fortunate enough to have had virtually all of their works preserved intact. With others, particularly the early Greeks, we have very little of their work preserved directly, and we have to use what are known as fragments – quotations from them, or descriptions of their views preserved in the work of later writers. The fragments of Xenophanes are very important. Firstly, Xenophanes was critical of popular religion, without being persecuted for his views. Secondly, his ideas illustrate some important features of the early Greeks. Critical of orthodox opinions, they had a self-awareness that allowed them to see a great deal of traditional religion as anthropomorphic. We see here also the cosmopolitan nature of the Greeks. As a trading and seafaring nation, they were aware of the views of other cultures, and able to take them into account.

Cosmos: an Elegant Universe

The ancient Greek word *cosmeo* has given us several words in modern English, such as 'cosmology' (the study of the universe), 'cosmogony' (the study of the origins of the cosmos) and 'cosmetic'. The last may seem somewhat surprising, but *cosmeo* not only meant to order or arrange, but also had a sense of good order; and also of beautiful, aesthetically pleasing order. A statement which is deceptively simple, but is in fact of enormous importance for the origins of science, is that the Greeks believed themselves to live in a *cosmos*, a well-ordered place. To them, the universe had an order, and a good and pleasing order at that. What is more, the Greeks were the first to recognise a distinction between the natural and the supernatural. They considered the cosmos to be an entirely natural place. Things did not happen at random, or by the caprice of the gods. With an optimism that was typical of them, the early Greeks believed the cosmos to be comprehensible. The order of the cosmos was something that could be discovered and understood by humans. Furthermore, they believed that the cosmos could be successfully described in words and numbers.

The first philosophers and scientists were the Milesians: Thales (*fl.* 585 BC), Anaximander (*fl.* 555 BC) and Anaximenes (*fl.* 525 BC). They came from Miletus in Asia Minor (on what is now the Turkish coast), an important cosmopolitan trading centre which had strong links with older Eastern cultures. Thales is said to have predicted an eclipse in 585 BC, to have been a brilliant geometer, and to have allowed a Greek army to cross a river by suggesting that they divert it into two streams, each of which was fordable. Anaximander was a pupil of Thales, and is said to have produced the first map of the known inhabited world. Anaximenes is reputed to have been a pupil of Anaximander. Unfortunately, little else is known about the lives of the Milesians, but they were the first to describe the cosmos in entirely natural terms, and the breadth of their theorising was quite remarkable. They also took the important step of not focusing on individual events (e.g., what caused this earthquake?), but looked instead at classes of events (e.g., what causes earthquakes?). This enabled them to form general theories about the causes of events, rather than specific theories about one event.

The philosopher Heraclitus of Ephesus (*fl.* 500 BC) insisted that the cosmos worked according to a *logos*, which in Greek could mean 'word', 'account', 'measure' or 'proportion'. We derive the word-ending '-ology' from *logos* (as in biology), as well as the word 'logo'. That the cosmos obeyed a *logos* meant not only that it was an orderly place, but also that it was comprehensible to humans if they could grasp the nature of this *logos*. The cosmos could be correctly described and understood using words. To do so, it was necessary for humans to generate a common account of the *logos*, not just individual accounts, and to follow it wherever it led. Thus, the Greeks were happy to pursue all sorts of arguments to their logical conclusion, which was an important factor in driving their theories forward. Heraclitus also said that:

This cosmos, the same for all, was made by neither god nor man, but was, is and always will be: an ever-living fire, kindling and extinguishing according to measure.

This expression of the idea of an objective, natural and orderly cosmos was typical of the early Greek scientists. Heraclitus believed that:

For those who are awake the cosmos is one and common, but those who sleep turn away each into a private world. We should not speak and act like sleeping men.

This is not to say that either the early Greek philosopher-scientists or the Greek populace were atheists. After all, the Greeks gave us a marvellous theology, full of characterful gods like Zeus and Poseidon. However, there was a huge difference between the sort of gods that societies prior to the Greeks – and the Greek populace – believed in, and the philosopher-scientists' conception of god. The popular gods had unpredictable wills of their own, and often interfered in human affairs and with the world in general, either deliberately or accidentally. The easy explanation for phenomena such as thunder and lightning was simply to say that they were caused by the gods, perhaps because they were angry with humans, perhaps because they were arguing among themselves.

The Milesians refused to use the gods to explain events in this manner. Events in their cosmos had natural causes. If the early scientists believed in gods, then theirs were gods who kept order and did not interfere with the natural processes of the cosmos. It was, of course, important that the philosopher-scientists should be free to have their own conception of god. Xenophanes, free to express his opinions, was not an atheist, but stressed the need for one god who behaved in a seemly manner and did not change his nature.

The history of ancient Greek science is the history of a relatively small group of people – far, far smaller than the scientific community of today. Their ideas were influential, both historically and in Greek society, but their views need to be distinguished from those of the Greek population as a whole, many of whom still held the old views about the gods and myths. Against that background, it is all the more remarkable that science began to come about at all. The views of the early philosophers were known and discussed in society, and *The Clouds*, a play by the famous Athenian dramatist Aristophanes (*c*. 450–385 BC), had philosophers debating whether lightning was caused by the clouds or by the gods.

Myths and Theories

There was another critically important distinction between the early Greek philosopher-scientists and their predecessors. This was that they sought to explain the cosmos in terms of theories rather than myths. Again, this is not to say that the Greek populace gave up believing in myths, or that the Greeks did not give us a rich and enduring mythology. Rather, a small group of people began to think in a different way and this was enormously important for the origins of science. The Babylonians and other societies prior to the Greeks generally used *mythopoeic* thought – thought expressed in terms of myths and poems. There were significant differences between myths and the sort of theories that the early Greeks were interested in. This allowed the Greeks to begin to make rapid progress in both philosophy and science.

Think of the different ways in which we compare myths and compare theories. How do we decide if one myth is better than another? There are criteria which might apply to myths, but not the same sort as apply to theories. Rather, they are literary criteria. Myths might be imaginative, entertaining, or carry some message by means of allegory, etc. So one might prefer the gods of Greek myth to the gods of Norse myth, or Tolkein's *The Hobbit* to *The Lord of the Rings*, on grounds of grandeur. Or vice versa. One might prefer simplicity to grandeur. These are subjective, rather than objective, criteria. Do we judge theories in this way? One might also consider this. There are many ways of telling the story of King Arthur. Is there a right way? If we stick to the historical evidence for Arthur, we get a pretty dull story. So we might embroider some mythology around him. But what do we embroider, and on what grounds?

We can collect evidence and discuss the merits of a theory. We have a good idea of why one theory is better than another. There is a need for theories to be consistent internally, or if we hold a group of theories, for them to be consistent as a group. Theories should also be as general as possible and have no exceptions. So there is a drive to establish a completely general theory. There is no great need for myths to be consistent with themselves or with other myths. Myths are compatible with one another in a way that theories are not. However, if I hold the theory that the sun is made entirely of stone, and you hold that it is made entirely of fire (an issue which concerned the ancient Greeks), then one of us is right and one of us is wrong. Theories exclude other theories from being true in a way in which myths do not.

Myths ask us to believe in a great number of things, while theories do not. There is an important principle applying to theories, which runs like this: *We do not suppose there to be more things in our theories than the bare minimum required to account for the phenomena.* This is known as the principle of parsimony, or Ockham's razor, after the medieval philosopher William of Ockham (*c*. 1285–1349). For example, we might have two ways of explaining why presents appear on

Christmas morning. Is there a Santa Claus with some magic reindeer who delivers presents down the chimney? A delightful piece of mythology, but the more mean-minded of us will say that humans have bought the presents and we have no need to believe in Santa and his reindeer. A good theory is very mean with what it supposes there to be. A myth, on the other hand, may be a good myth because it has more, or more extravagant, magical monsters. The problem is, what criteria do we have for determining how many and what sort of monsters? Ultimately, the difference between myth and theory is this. A good myth may be no nearer the truth, in terms of being a description of the world, than a poor myth. A good theory is.

Theories can be debated in a way in which myths cannot, because of the objective criteria for theories. We can collect the evidence, debate the merits of competing theories and agree on the best theory. It is by no means clear that one can do something similar with myths. I have no wish to dismiss myths, which are fine examples of human creativity, imagination and ingenuity. Myths have their place, but they also have their limitations. One important limitation is that myths do not generate or drive progress in the way that theories do. The requirements that theories be consistent, that they cover all the evidence and be as general as possible while being as parsimonious as possible, mean that it is clear what form a better theory might take.

The Greek use of theory can be considered alongside their idea of the cosmos as a well-ordered, natural place. The cosmos was to be explained in terms of theories which used natural explanations only. As Heraclitus insisted, there was a need for an objective account of it. There was a belief that its nature could be described and comprehended by human beings using words and numbers. Once the Greeks formulated the idea of a cosmos, and began to try to explain it in terms of theories, then their science and their philosophy began to progress very rapidly. They made leaps of sophistication in their theories of matter and their cosmology, and many other sciences appeared which were simply not seen in other ancient societies.

Natural Phenomena

Do we know that the Babylonians and Egyptians had no theories? We have a considerable amount of evidence about Babylonian astronomy, but we have yet to find any real evidence of a proper theory of the heavens. The ancient Greeks, and in particular Aristotle, were fascinated by theories and wrote down every one they could find. Aristotle's aim was usually to show these theories to be wrong and himself to be right. If he had known of any Babylonian theory, it is very likely that he would have recorded and criticised it. He was certainly aware of Babylonian and Egyptian culture, and indeed praised the accuracy of their astronomical observations. Yet he tells us that:

Others state that the earth rests on water. This is the most ancient account we have, which was given by Thales of Miletus, that it stays in place through floating like a piece of wood or something similar . . . as if the same argument did not apply concerning the water supporting the earth as to the earth itself.

The point here is not that this theory is right, nor even that it is a particularly good theory. It is that it attempts to explain the world in a certain manner, without resort to myth or poetry. It makes a clear statement about the nature of the earth which can be discussed and verified or falsified. It is a theory, depending on natural things only, and not a myth. It can be sharply contrasted with the Greek myth that the world was supported on the shoulders of Atlas. So too with many other early Greek theories. The later chronicler Aetius tells us that:

Concerning thunder, lightning, thunderbolts, whirlwinds and typhoons, Anaximander states that all these come about because of wind. Whenever it is enclosed in a thick cloud and then forcibly breaks out, due to its fineness and lightness, then the bursting makes the noise, and the rent against the blackness of the cloud is the lightning flash.

Again, this may not be the greatest of theories, but it attempts a natural explanation of phenomena that had previously been explained by the actions of the gods. Zeus was supposedly responsible for thunder, Poseidon for earthquakes.

So the Greeks were responsible for a new conception of nature as a cosmos, and indeed for the very distinction between natural and supernatural. It could even be said that they 'discovered' nature. So, too, they were responsible for producing the first theories, devoid of supernatural influences, of how nature might work. They effected the transition from myth to theory. Next we shall see the rapid progress they made with their theories and their idea of a cosmos, as they sought to carry through their vision of a new way to explain the world.

2 The First Scientific Theories

Which was to be proved.

Euclid, *Elementa*, book 1, proposition 5 and *passim* (usually quoted in Latin: '*Quod erat demonstrandum*')

The first ideas about the nature of the world gave a hemispherical picture of the universe. There was the earth, which was flat, and a hemispherical heaven above it. There was no conception of anything beneath the earth (such as the other half of the heavens), and so no question of whether the earth needed some form of support to hold it in position. Thales made a conceptual leap forward in seeing the problem of what might lie below the earth. He gave us the first spherical, as opposed to hemispherical, cosmos. Aristotle's criticism, doubtless voiced by many others before him, was that water is something we know to be heavy, so why does it not fall as well? The next generation of cosmologies had the earth supported by air, something which in our experience does not fall. Aristotle tells us that:

Anaximenes, Anaxagoras and Democritus say that it is the flatness of the earth that is the reason why it stays still, for it does not cut the air below but covers it as flat bodies are able to do.

So the earth rides on air, a bit like a frisbee. For us today, these theories seem slightly odd, but the resources available to the Greeks were very limited. The following statement may seem surprising, but it can explain a great deal about the nature of Greek cosmology and cosmogony: the Greeks had no conception of gravity. Of course, they knew that if you released a heavy object it fell to the ground. They had no idea, though, of bodies having a mutual gravitational attraction, or of an object falling because of the gravitational attraction of the earth. Instead, they had to explain such phenomena in other ways. So early cosmologies had a direction to the cosmos, a top and a bottom, with heavy objects falling from one to the other. Hence the problem of why the earth appeared not to be falling. The Greeks began to realise the problems, though, and moved to a further stage of sophistication. Aristotle tells us that:

There are some, such as Anaximander among the ancients, who say that the earth rests on account of its likeness. It is fitting that what is established in the centre and has equal relations to the extremes should not move up, down or to the side. It is not possible for it to move in opposite directions at the same time, and so necessarily it remains still.

So the earth was central and immobile, and would remain so without any means of support. Furthermore, the earth was believed to be spherical. From this time on, it was the view of every educated person in ancient Greece that the earth was a sphere, and not flat. They supported this view with observations. They saw the shape of the earth's shadow cast on the moon during an eclipse, and were aware that if you journeyed much further south than Greece, different stars could be seen in the night sky. At sea, the appearance of ships over the horizon was explained by a spherical earth. The Greeks explained the fact that we stick to this spherical earth by means of a 'like-to-like' principle. They held that like things attract, and since we are made of the same sort of stuff as the earth (a radical thought in itself), we stick to it.

The Greeks also began to have theories of cosmogony, and of the origins of life. Early mythical cosmogonies were often of a sexual nature – typically, the earth might come about due to the sexual coupling of two gods, perhaps of sky and sea. According to Anaximenes:

The stars came about from earth, through the moisture rising from it, which when rarefied becomes fire, and the stars are composed of fire which has been raised high.

He went on to say that:

> *Living things were generated by water being evaporated by the sun. Humans, in the beginning, were similar to another animal, namely to fish.*

For the first time, we see an account of the origins of the cosmos and of life in entirely natural terms. Crucially, mankind was located firmly in nature. Once the Greeks began to have theories about the nature of the heavens and the earth, rather than myths and poems, the sophistication of their cosmologies rapidly increased. We do not see anything like this in ancient societies which had a dependence on mythopoeic thought. A civilisation as technologically advanced as the Egyptians, the architects of the great pyramids, believed that the goddess of the heavens, Nut, formed a hemisphere over the god of the earth, Qub (see Figure 1).

In addition to this, the Greeks began to develop the first theories of matter. This began with Thales, whose theory was that everything was, at root, water. Whether he meant that everything was now water in some form or another, or that everything to begin with had been water (and that some had transformed into something else), is unclear. What is clear, though, is that Thales had a conception of matter as entirely natural. His fellow Milesians Anaximander and Anaximenes agreed that there was only one substance. Anaximander opted for

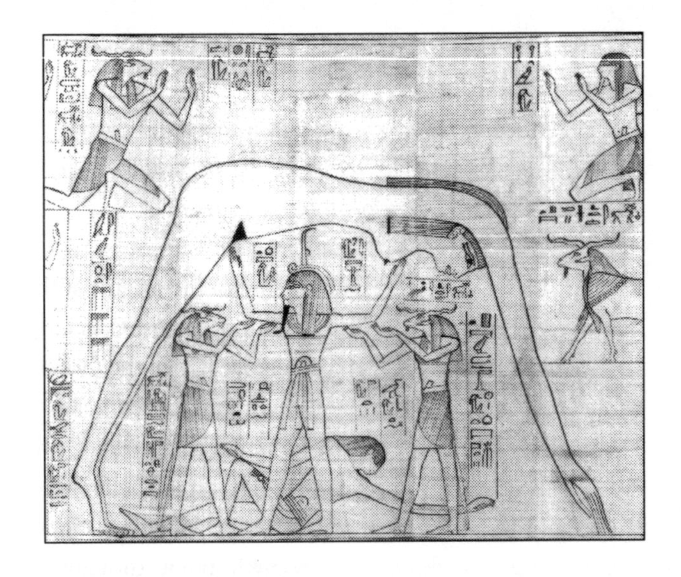

Figure 1: From the Egyptian Book of the Dead, *c.* 1000 BC. (Source: British Museum)

'the unlimited', and Anaximenes for air. The problem with Thales' theory, as far as the Greeks were concerned, was this: how could water become fire? The point of Anaximander's 'unlimited' may have been that the unlimited could become anything, while Anaximenes was keen to argue that air could change by condensation, becoming first water and then earth, or could become fire by rarefaction, and that all these processes could be reversed. Later, Heraclitus, who believed that all things were in a process of permanent change governed by a *logos* (hence his famous assertion that 'you cannot step in the same river twice'), suggested that the one substance was fire.

The philosopher Empedocles of Acragas (492–432 BC) introduced the four elements of earth, water, air and fire, which were to become standard in Greek thought. These should not be understood as literally earth, water, and so on, but in a slightly more abstract manner as solid, liquid, gas and fire, or as principles of solidity, fluidity, gaseousness and fieriness. Earth, water, air and fire had always been a Greek classification of matter, but it was Empedocles who took the bold step of insisting that these were the four basic elements. He also insisted that physical objects were fixed proportions of these elements. For example, metals would be fixed proportions of earth and water (the Greeks considered metals to be made partly of water, since they became fluid when heated). As with cosmology, once the Greeks began to theorise, there was a rapid increase in the sophistication of their theories of matter.

Despite the almost universal acceptance of the four elements after Empedocles, there were wide variations on exactly what constituted them, as we will see with Aristotle and Plato. It was not until the chemical revolution of the eighteenth century that air was shown to be composed of several different gases and that water could be made from, and dissociated into, hydrogen and oxygen, thus showing that air and water were not elements. It was not until the nineteenth century that the idea that fire, or heat, was some form of substance was abandoned.

The Fathers of Medicine

There is no doubting that the Babylonians, and indeed the Egyptians, had some effective healing practices and a reasonable knowledge of the human body. However good their healing techniques, though, the Babylonians did not consider health and disease to have physical causes. Rather, they saw disease as a punishment from the gods for some sin committed. Thus the first task of the doctor was to diagnose the sin, and then to work out a means of purification to absolve it. The Babylonian approach was in very sharp contrast to that of the Hippocratics. We know relatively little of the life of Hippocrates of Cos (*c*. 460–370 BC), but we do have an extensive set of writings known as the 'Hippocratic corpus', composed by Hippocrates and his followers between 430 and 330 BC. They believed that every disease had a physical cause, and that no disease was caused by the intervention of the gods. The Hippocratics were very forthright about this, to the extent of setting out their views on the most difficult sort of disease for them – epilepsy, known to the Greeks as the 'sacred disease'. It was commonly thought that epileptic fits were due to possession by the gods, to some divine intervention, and hence 'sacred'. Yet the Hippocratics put their point bluntly and forcefully in the opening passage of *On the Sacred Disease*:

> *In my opinion, the so-called 'sacred disease' is no more divine or sacred than any other disease, but has a natural cause, and men consider it divine because of inexperience and wonder, it being unlike other diseases.*

In *The Science of Medicine*, their view is crystal clear:

> *Every disease has a natural cause, and no event occurs without a natural cause.*

On the Sacred Disease goes on to say:

> *The so-called 'sacred disease' comes from the same causes as the rest . . . each disease has a nature and power of its own, and none is unintelligible or untreatable . . . whoever knows how to bring about moistness, dryness, hotness or coldness in men can cure this disease as well, if he can diagnose how to bring these together properly, and he has no need of purifications and magic.*

Again, all diseases have a cause. This is a quite remarkable piece of optimism, typical of the early Greek thinkers. There is nothing in any disease which is 'unintelligible or untreatable'. There is also an attack on magic in general. Prior to this, there had been criticism of individual magicians for being incompetent at their art, but with the early Greeks we find something quite new – the first recorded attack on magic as such. Magic and the supernatural simply did not exist. The world was a purely natural place, to be explained by natural means. There were not competent and incompetent magicians – there was simply no such thing as magic, and so all magicians were charlatans. The Hippocratics attempted to back up their views on the sacred disease through the use of experiments and reasoning. They opened the heads of goats, which suffer from a similar disease, and in *On the Sacred Disease* we are told that:

> *If you cut open the head you will discover the brain to be moist, full of fluid and rank, showing that it is not a god who is harming the body, but the disease.*

Here was the physical basis of the disease. They criticised the 'purifications' which involved diet (because of the similarity of the disease in goats, people were advised not to eat goat meat or use goat skins). If one could avoid epilepsy by diet, then the disease could not be supernatural, said the Hippocratics. This insistence on the physical nature of all diseases, and the assault on magic and the supernatural, was a typical aspect of early Greek thought, and crucial in separating Greek science from that which preceded it.

Eleatics and Atomists: Achilles and the Tortoise

The Greeks also began to consider some more abstract scientific questions. The Eleatic philosophers (from Elea, a Greek colony in southern Italy), Parmenides (*fl.* 480 BC) and Zeno (*fl.* 445 BC), investigated certain problems concerning motion and change at a highly abstract and theoretical level. The problems that they discovered were influential in the development of Greek science and philosophy. Zeno is most famous for creating a series of paradoxes, one of which has come down to us in the form of 'the tortoise and the hare'. If a tortoise has a head start in a race, can the hare ever catch him up? Each time the hare runs half the distance to the tortoise, the tortoise will have moved on a small amount – so can the hare ever catch him? A different version of this paradox seems to show that it is impossible to move at all. All motion takes a certain amount of time, however small, you will agree. But if I try to walk to the door, first I must go half way, then half of what is left, then half of what is left again, *ad infinitum*. If each of these motions takes an amount of time, however small, and there are an infinite number of them, then it will take an infinite amount of time to complete my journey. So I can never reach the door! What is worse, I cannot even start, since the first part of my journey is infinitely divisible in the same manner.

Zeno had a whole bag full of arguments such as these.

While his paradoxes may seem frivolous, they played on important questions of divisibility and infinity. The central issues of concern for him were whether space, time and matter were continuous, and so infinitely divisible, or whether they were discrete – that is, occurring in small indivisible packages.

Parmenides, too, was worried about the nature of change. His worry was deceptively simple, and can be put into two seemingly innocuous statements: what is, is; nothing comes from nothing. But if that is so, how can anything ever change? What is, cannot change, nor can what is not; nor can anything come to be from what is not. Yet this seems highly paradoxical, as do Zeno's ideas, for there seems to be change all around us. Why did Parmenides and Zeno produce these paradoxes? At least part of the reason must be the Greek attitude to *logos*, which we saw in the last chapter. The Greeks were willing to follow an argument to the end, no matter what the consequences.

Both Parmenides and Zeno produced paradoxical conclusions which stimulated others to delve much further into these matters. If the cosmos was comprehensible, then Parmenides and Zeno had to be wrong. The most important solution to the problems of change posed by the Eleatics came from the first Atomists, Leucippus of Miletus (*fl.* 435 BC) and Democritus of Abdera (*fl.* 410 BC). It is likely that Leucippus invented atomism and that Democritus refined it, but the sources tend not to distinguish between their opinions or their roles. Their view was that there was a void or vacuum in which there were discrete particles of matter. These particles were atoms, from the Greek *atomos*, meaning 'uncuttable'. Unlike modern atoms, these particles were indivisible and did not undergo any change in themselves. However, these particles did move around in the void, becoming arranged in different patterns. These altered configurations of the atoms were perceived by humans as change. Ancient atomism is important for two reasons. Firstly, it is the ancestor of modern atomism. Though the modern theory of the elements is far more sophisticated, ultimately its roots are with Leucippus and Democritus. Secondly, ancient atomism was also the first properly thought out two-level theory of the world. It distinguished between what human beings *perceive* and how the world actually *is* at the atomic level. It introduced the idea that the reality behind appearances might be radically different from those appearances. Democritus said that what we perceive is:

> *By convention sweet, by convention bitter, by convention hot, by convention cold, by convention colour, but in reality atoms and the void.*

Why did the problems posed by Parmenides and Zeno worry the Greeks so much? We come back to the idea of theories, and Greek assumptions about the cosmos. The

Greeks assumed the cosmos to be comprehensible, to be understandable by humans. If so, then they required a coherent theory of change, and the paradoxes raised by Parmenides and Zeno had to be resolved. If Parmenides and Zeno were correct, then everyone else was wrong. Their views were not compatible with other theories, and had to be overcome. This requirement drove the Greeks to new heights of sophistication in their conceptions of matter and change.

The Pythagoreans: the Secret Magic of Numbers

Also of considerable importance were the philosopher and guru Pythagoras of Samos (*fl.* 525 BC) and his followers. The Pythagorean brotherhood was both secretive and religious, and held odd views about reincarnation and the transmigration of souls. Pythagoras is said to have stopped a man beating a dog, saying that he heard the voice of a reincarnated friend in the bark of the dog. Pythagoras himself may well have travelled in Egypt and learnt Egyptian geometry; certainly, some of his religious practices (such as not eating beans) show Egyptian influences. Pythagoras founded a school of religion and philosophy at Croton, in southern Italy, and his students devoted themselves to mathematics and spiritual purity, renouncing personal possessions and following a vegetarian diet.

The Pythagoreans made some important contributions

to mathematics and science. They took mathematics and geometry beyond the practical stage and developed them theoretically. It may seem odd to talk of practical geometry, but it is due to the Greeks that we have geometry in its current form. An example of practical geometry would be as follows. How can you measure the height of a tree? Cut a stick to your own height. When your shadow is as long as this stick, measure the length of the tree's shadow. Geometry literally meant 'land measuring' (*ge*, earth, and *metreo*, to measure), and many pieces of practical geometry were based on measuring land in agriculture. It was typical of the Greeks that they sought to convert these practical procedures into a precise and abstract science. They converted the technology of practical geometry into the science that we know today. Pythagoras' theorem about right-angled triangles was known well before this time, but was now proven. Another Pythagorean called Archytas of Tarentum (*fl.* 385 BC) solved the theoretical problem of how to construct a cube of twice the volume of the original.

The Pythagoreans were also the first to think about the relationship between mathematics and nature, investigating numerical relationships in acoustics and musical theory. They discovered that if you halve the length of a string, then you obtain a note an octave higher. They also discovered that other notes could be expressed in terms of mathematical ratios, such as a

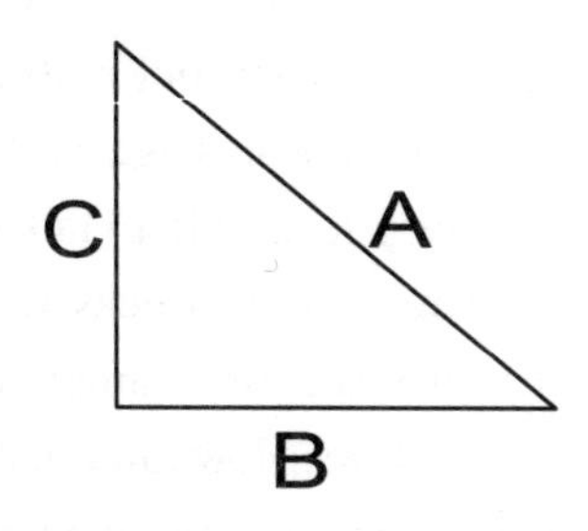

Figure 2: Pythagoras' theorem. For all right-angled triangles, the square of the length of the longest side equals the squares of the lengths of the other two sides added together, so $A^2 = B^2 + C^2$, or $A = \sqrt{(B^2 + C^2)}$

fifth and a fourth being 1:2 and 3:4 respectively. The discovery that numbers could describe the world so well fascinated the Pythagoreans, and is their most important legacy. At the outset of the scientific revolution of the seventeenth century, Galileo declared that:

> *Philosophy is written in a huge book, which stands open before our eyes, that is the universe. It cannot be read until we have learnt the language and become accustomed to the symbols in which it is written. It is in a mathematical language, and the letters are triangles, circles and other geometrical figures, without which it is not possible to understand a word.*

Nowadays, physicists do indeed believe that the world has a deep mathematical structure, best expressed in terms of mathematically framed laws of nature such as $f = ma$ or $e = mc^2$. The Pythagoreans, as pioneers of the idea that numbers are important in science, were understandably a little more naïve about the relation between numbers, nature and science. In fact, they went rather overboard, seeing numbers and harmony everywhere. Most famously, the Pythagoreans believed in a harmony of the heavens. The motions of the stars and planets were so related mathematically that a harmony was produced, though it was inaudible because we had heard it from birth and it was now background noise. They also believed in many seemingly arbitrary relationships, such as the male number being 3, the female 2 and the number of marriage 5. They even believed that the world was in some way constructed from numbers, rather than matter.

We should not be too harsh on the Pythagoreans. It was not until the scientific revolution of the seventeenth century that something like the modern notion of mathematical laws of nature was worked out, and they did introduce the important idea that the cosmos could be captured in numbers as well as words, and that it was amenable to precise mathematical description. The Pythagoreans had one very nasty shock, though. They discovered that $\sqrt{2}$ is not a rational number. That is, the square root of 2 cannot be expressed as the ratio of two whole numbers, as a:b. This was deeply disturbing to the Pythagoreans, who believed numbers to be part of the fabric of the world, and that all numbers were rational. Later legend has it that the Pythagorean who revealed this secret was drowned by divine retribution.

Come the Time of Proof

We have seen that once the Greeks began to formulate theories, their ideas developed rapidly. They also began to investigate some more abstract questions about the nature of the world they lived in. Along with this came the development of ideas about the nature of argument. This is in many ways characteristic of early Greek thinkers, who loved such abstraction. The problem they faced was this. Once you have theories which are not compatible with one another, then you need a procedure to decide between those theories. Certainly, the Greeks loved their philosophical debates, and theories were put to the test in discussions between philosophers. The quality of the theory and the evidence for it were examined minutely. However, some of the Greeks came to realise that there was another factor in these debates apart from the strength of the theory, which was the ability of the debater. It worried the ancient Greeks that the weaker theory could be made to defeat the stronger, in the hands of a good debater. There also emerged a group known as the Sophists, who were professional arguers. They would take either side of a debate, depending on who paid them and what the audience wanted to hear. Plato in particular was savage about the Sophists, insisting that the goal of philosophical argument should be truth, and not just defeating your opponent. Plato was keen to distinguish rhetoric – devices for winning over a crowd – from good philosophical argument.

The Greeks began the development of formal logic. We can find the first intimations of this in Plato, and the first systematisation in Aristotle. Logic is frequently misunderstood. It is about the form, rather than the content, of arguments. It tries to decide which forms of argument are valid (and therefore good arguments) and which are invalid (bad arguments). Here is a simple example. Take the argument that 'if all humans are primates, and all primates are mammals, then all humans are mammals'. Does this structure of argument work generally? In other words, if all A's are B, and all B's are C, are all A's C? If the answer to that is 'yes' (and it is), then this is a valid argument, and it works whatever A, B and C are, as long as all A's are B and all B's are C. Can we also therefore conclude that all C's are A (in our example, that all animals are humans)? If the answer is no (and it is) then this second argument is invalid and ought not to be used. Logic studies the validity of arguments at this general level. Although this is a simple case, logic can be very helpful in working out whether or not more complex arguments are good ones, and so deciding between theories.

Euclid

We can also find the Greeks doing something very important with geometry, and here the key name is Euclid (*fl.* 300 BC). Euclid attempted to give geometry stable foundations and a thorough, rigorous structure. He began with definitions ('a point is position without magnitude, a line length without breadth', etc.), postulates (assumptions about points and lines) and axioms (assumptions needed to generate proofs, such as 'things which are equal to the same thing are equal to each other'). He then proceeded to prove, in an absolutely rigorous manner, a great number of geometrical theorems. Some of these were relatively straightforward, but he also managed to prove many complex theorems and discovered a great deal about the structure of geometry. The great beauty of Euclid's geometry was that if you agreed with the definitions, postulates and axioms, then the proofs compelled you to believe the more complex theorems. For one science at least, here was a definitive procedure for resolving disputes and making progress.

Attempts were made to ensure that Euclid's system became even more rigorous, by reducing the number of assumptions (definitions, postulates, axioms) which were not proved but which had to be accepted. The fewer assumptions, the more definitive the system. Attention centred on Euclid's fifth postulate.

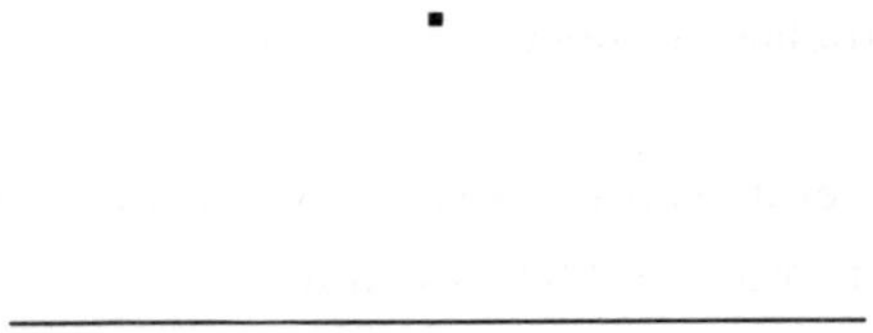

Figure 3: Euclid's fifth postulate. Given a straight line and a point on a plane, how many lines can be drawn through that point parallel to the line? Euclid's answer was one, and only one.

Here we have a point and a straight line on a plane. The question is: how many straight lines can be drawn through the point which are parallel to the line that we already have? Possible answers are 0, 1 or more than 1. The intuitive answer, and the one assumed by Euclid, is that there is one, and only one, parallel line. It was felt by many geometers that this postulate ought to be provable from the other postulates, definitions and axioms. Attempts to do so failed, however. So, too, did attempts at indirect proof. Indirect proof takes the alternative answers to the one that you want to prove, and attempts to show that they lead to contradictions. If all of the alternatives lead to contradictions, then you have indirectly proved your answer. So people attempted to show that the condition of no parallel lines, or of more than one parallel line, led to a contradiction, but they failed to do so. There is an interesting Greek assumption here. The world is comprehensible, so there must be a non-contradictory way of describing it. If you rule out all of the contradictory answers, what you are left with must be right!

It was eventually realised, in the nineteenth century, that there was a subtle assumption about the nature of space involved here. Euclid's geometry, with one parallel line, described flat, homogeneous space (the sort of space we are familiar with from a flat piece of graph paper). A geometry with no parallel lines was self-consistent, and described curved surfaces like that of a sphere. Think of a globe. All lines of longitude are at 180° to each other at the equator but all meet at the poles. This is a positive curvature. With a negative curvature (such as a saddle), in which lines run away from each other, there may be many lines drawn through a point which never meet. These alternative geometries were of only theoretical interest until the advent of Einstein's theory of general relativity, when it was realised that space might have a positive or a negative curvature. Euclid's geometry is a major achievement, and remains valid for Euclidean (non-curved) space. It is also an important expression of the way in which the Greeks sought both secure starting points for their science and ways of resolving disputes over their theories.

Science and Technology

A society in possession of science should be conscious of a distinction between science and technology. We find no such distinction in any of the pre-Greek societies. The Greeks, though, distinguished between *empeiria*, meaning a knack or skill acquired through practice, and *episteme*, meaning knowledge, which required the ability to give reasons why something was the case. The person having *empeiria* might be able to manipulate the world, but he would not be able to explain why what he was doing should work. A typical example for the Greeks was the difference between someone who knew a few folk remedies for disease, and a doctor who knew the nature of the body and could explain why, how, and in what circumstances those folk remedies would be effective. Plato in particular was keen on this distinction. He made a contrast between the capabilities of someone with a basic empirical or practical acquaintance with a subject, and the theoretical and synoptic knowledge that an expert might be expected to have. The Greeks were conscious of doing something different from – and more sophisticated than – technology, and were also aware that their attitudes to myth and religion were different from those of other societies.

Great Achievement Assured

Let us sum up the achievements of the early Greeks. Their society was rather different from other ancient societies, not being a hierarchy or having a centralised state religion. There was a greater tolerance of debate, and also an affluence which allowed well-to-do people the time to contemplate philosophical and scientific questions. The Milesians began to think about the world in a radically different way. They believed their world to be a cosmos, a well-ordered entity, which could be understood and explained by people using words and numbers. They distinguished between the natural and the supernatural, believing the cosmos to be entirely natural and subject to regular behaviour. Gone was any notion of mythical explanation. The Milesians dealt in theories, and tried to improve on each other's theories. They produced the first theories of matter and the first cosmologies and cosmogonies that are couched in natural terms. Their theories may appear naïve to modern eyes, but that is only to be expected of thinkers who were only just beginning to produce scientific theories. The key point is that they were theories rather than myths.

The Hippocratics rejected the idea that diseases are punishments sent by the gods, asserting that all diseases have a physical cause. They also produced the first criticism of magical practices on a general level. We also see the Greeks begin to pursue more theoretical questions in science. The Eleatics posed problems about change and motion, and whether space, time and matter are continuous or atomistic. The atomists answered that matter, at least, comes in small, unsplittable packages called atoms, and so developed the first two-level theory of appearance and reality. We also see the development of mathematics and geometry beyond its practical phase. With the Pythagoreans, we find the first investigation of the way in which mathematics may be linked to physics, and the first attempts to prove geometrical theorems. The Greeks became interested in how to resolve debates about their theories, introducing notions of demonstration and proof. They were also conscious of a distinction between science and technology. However crude the initial theories of the Greeks, they were scientific theories as opposed to myths, and from there the Greeks made rapid progress to more sophisticated theories, something unseen in other ancient cultures. The Greeks carried through their vision of a new way to explain and investigate the world with exemplary thoroughness and enthusiasm.

3 Men of the World

Nature does nothing without purpose or uselessly.
Aristotle, *Politics*, book 1, 1256b, 20–1

The great majority of the ancient Greeks whom we have met so far are known as 'pre-Socratics'. The history of Greek philosophy divides into three periods: the pre-Socratics, who worked before Socrates (469–399 BC); the period of the three great Athenian philosophers, Socrates, Plato (427–348 BC) and Aristotle (384–322 BC); and the period of the Hellenistic philosophers, who all worked after Aristotle. Socrates, though he was immensely important in the history of philosophy, was more interested in ethics than science. Plato and Aristotle are of great significance, though. Aristotle's views came to dominate both the ancient world and Western thought in general until the scientific revolution of the seventeenth century. Even thinkers who were not specifically Aristotelians borrowed many of his ideas. His views were later synthesised with Christian theology to form a philosophy known as 'scholasticism', and it was this that the scientific revolution sought to transform.

Plato's views, while sharing a good deal of ground with Aristotle, were a consistent alternative to Aristotelian ideas. Plato's thinking underwent a revival in the form of 'neo-Platonism', both in the later ancient world and during the Renaissance.

Athens in this period formed the intellectual hub of Greece, not only in philosophy and science but in the arts and politics as well. Politically, it was a melting pot, veering rapidly from the first experiments in democracy, to various forms of oligarchy (rule by the few), to dictatorship. Athens was also frequently at war with other city states, most notably Sparta. Freedom of expression, and a spirit of open debate and criticism, allowed philosophy, science and the arts to flourish. Theatre and poetry prospered, great rhetorical speeches were made on the issues of the day, fine buildings were put up, and philosophers debated openly, often to audiences. Philosophical symposia were held, in which the great philosophical issues were discussed. These could be polite dinner parties, or rather wild drinking sessions. Plato's book *The Symposium* records one of these events, and discusses the nature of love. There were representatives of many philosophical viewpoints on all the major issues of Greek thought. In relation to science, the main views were those of the atomists, Leucippus and Democritus, and of Empedocles and Anaxagoras, and of the Milesians, Thales, Anaximander and Anaximenes. Plato tended to group these people together, calling them the '*physiologoi*', those who explained in terms of nature. There were also the views of the Eleatics, Parmenides and Zeno. Plato and Aristotle built on the work of these philosophers in interesting new ways, and their thinking was seminal in forming the philosophical and scientific legacy of the Greeks.

Plato

Plato was born in 427 BC, probably in Athens, and in his youth was a devoted pupil of Socrates. Socrates professed to know nothing himself, but was remarkably good at conducting philosophical investigations. He was brilliant at showing pompous people who thought that they had definitive answers to philosophical questions that in fact they did not – that they needed to think more deeply. A typical Socratic question was: 'What is courage?' Socrates would then refute other people's ideas about courage without giving a definition himself, prompting people to rethink their views. No respecter of dignity or position, Socrates got himself into trouble. The death of Socrates – he was executed by the Athenians on trumped-up charges of impiety while defending his philosophical views – affected Plato deeply. Plato devoted himself to philosophy, and remains one of the truly great philosophers.

Plato had an eventful life. He served as a soldier in some of Athens' wars and was often embroiled in political matters. He became disenchanted with Athenian politics and politicians, and argued that until philosophers became rulers, or rulers philosophers, things would never go well for the state. Later in his life, he attempted to put some of these ideas into practice. He travelled to Syracuse in order to educate the young Dionysus II, in the hope of making him a philosopher-king. However, after intrigue and betrayal the project failed and Plato barely escaped with his life. In Athens, he founded the Academy, a school and research group for philosophers, which was the first of its type.

Plato's philosophical writings are remarkable for their artistic as well as their philosophical merit. He wrote his works as dialogues, discussions of philosophical questions between the characters. Often the main protagonist is Socrates, and there is much academic debate about how far Plato's works record actual conversations and how far they are fictional. Whatever the answer to that question, Plato gives us a unique insight into the sorts of debate that were going on in Athens. He has a remarkable ability to make these debates come alive, and a notable talent for the characterisation of the protagonists. These characters are often related to the philosophical position that they are given to argue, and the dialogues are laced with humour and dramatic turns to the debate. Plato's works are literature and philosophy at the same time, something very rare in philosophy.

Plato was more important for his attitude to the investigation and conception of the world than for any specific theories. His steadfast belief was that the order of the cosmos could not have come about by chance. Plato was keen to contrast the potential chaos of the world with the apparent good order of the cosmos. The order of the heavens, the beauty of the world around us, the nature of living beings, all stood in stark contrast to a possible chaos in which the elements were randomly distributed. He also imagined a chaos in which matter itself was not organised, so that there were no recognisable elements like earth or water, just a mess. There was still no conception of gravity, but Plato and others believed in a like-to-like principle whereby like things were brought together. This was not an attractive force as such, more a principle by which things were sorted into groups from an initial chaos. Plato saw that such a principle would only sort similar things, and would not produce the order of the cosmos. Characteristic of the cosmos, according to Plato, was the proper proportion of *unlike* things together. Talking about his opponents' views, he said:

> *Let me put it more clearly. Fire, water, earth and air all exist due to nature and chance they say, and none to skill, and the bodies which come after these, earth, sun, moon and stars, came into being because of these entirely soulless entities. Each being moved by*

chance, according to the power each has, they somehow fell together in a fitting and harmonious manner, hot with cold or dry with moist or hard with soft, all of the forced blendings happening by the mixing of opposites according to chance. In this way and by these means the heavens and all that pertains to them have come into being and all of the animals and plants, all of the seasons having been created from these things, not by intelligence, they say, nor by some god nor some skill, but as we say, through nature and chance.

Opposites such as hot and cold could not have been brought together by a like-to-like principle, nor was it likely that they would come together in a harmonious manner by chance. Some god or skill must have achieved this. This is an early example of an 'argument from design'. If we look at the workings of a mechanical watch, they appear too complex and well organised, too clearly arranged for a purpose, to have come about by chance. Therefore we believe there to have been a watchmaker. Design arguments state that the world is similarly complex, organised and arranged, so that we must suspect the presence of a designer. Plato believed that a god organised the cosmos out of primordial chaos. The nature of this god is of considerable importance. Plato conceived of him as a skilled craftsman, a *demiourgos*. It was because the god was skilled that he could form an organised cosmos out of chaos. When he did so, he did the best job possible, taking into account the limitations of his materials. Matter acted by necessity, similar things grouping together, and did not produce what reason and intelligence would produce. So Plato speaks of reason being imposed upon necessity, as far as possible, to produce the order of our world.

Plato's *demiourgos* was without jealousy and malice, and acted to create the best cosmos that he possibly could. Plato's conception of his god was radically different from previous Greek notions of the gods. The gods of Greek mythology were, in short, a pretty rough lot. They schemed, they plotted, they committed adultery and murder and generally acted in an amoral and unpredictable manner. As Plato complained, they seemed to have all the moral shortcomings of human beings. Plato's *demiourgos* acted only in the best manner possible. He did not interfere with the world once it was set up.

The *demiourgos* was also a geometrical god. That is, he employed the principles of geometry and mathematics to give the best possible proportion and harmony to the world and its constituents. Specifically, he imposed shape and number on the primordial chaos in order to create the cosmos. The world, then, had an underlying geometrical and mathematical structure. There was a purity and timelessness about mathematics and geometry which appealed to Plato. Truths about the physical world are subject to change (today it is 20°C, yesterday it was not); the truths of mathematics and geometry are not. Since Plato believed that knowledge should be indisputable and unchanging, mathematics and geometry were for him perfect examples of what knowledge should be like. Because his god constructed the world using geometry and mathematics, we should also be able to use them in order to understand the structure of that world. Likewise, because the *demiourgos* has created the best world possible – simple, elegant and aesthetically pleasing – our explanations should be simple, elegant and aesthetically pleasing.

Teleology: the Best of all Possible Worlds

The fact that the *demiourgos* created everything in the best possible manner had further important consequences for the way in which Plato explained the world. Plato introduced a type of explanation known as 'teleological'. This is, literally, an 'end-directed' explanation (from the Greek *telos*, meaning 'end'). The end that the *demiourgos* had in mind was to produce the best cosmos possible. Everything in that cosmos was designed to fulfil that purpose. Therefore, according to Plato, we can explain aspects of the cosmos teleologically by saying: 'It is so because it is best for it to be so.' Plato believed the stars to have regular and circular motion, because that was the best motion for them to have. Modern physics has done away with teleology entirely. We do not think that there is anything intrinsically 'good' about the arrangement of the universe. It just is as it is, and we explain how it works mathematically. The ancients used teleology because they could not see how anything as good (useful, beautiful, well organised, etc.) as animals, humans and the cosmos could come about through blind chance and the like-to-like principle.

Teleology pervaded a great deal of ancient thought, and in many ways dominated thinking about the cosmos right up to the scientific revolution. It was an aspect of Greek science always open to criticism, especially as some Greeks, notably the atomists, attempted explanations without the use of teleology. While we can plausibly explain the origins of order in the cosmos and the formation and development of living things without resorting to teleology, the ancients could not. The Greeks had good reason to be suspicious of the theories of the atomists on these matters, because with the resources available to the ancient Greeks, they seemed highly implausible.

Atomism: Let us Trace the Pattern

We can see a great number of the basic principles of Plato's world view in action in his atomic theory. Plato did not believe that the standard Greek elements of earth, water, air and fire were the ultimate building blocks of nature. He believed that these elements had a specific structure which could come apart, and that there were two basic entities out of which the elements were made. These were two types of triangle (see Figure 4).

These triangles did not undergo any change in themselves, and did not break down into anything more fundamental. These, for Plato, were the basic building blocks of nature. Why?

> *Of the two triangles the isosceles has one nature, the scalene an unlimited number. Of this unlimited number we must select the best, if we intend to begin in the proper manner. If someone has singled out anything better for the construction of these bodies, his victory will be that of a friend rather than an enemy. We shall pass over the many and postulate the best triangles.*

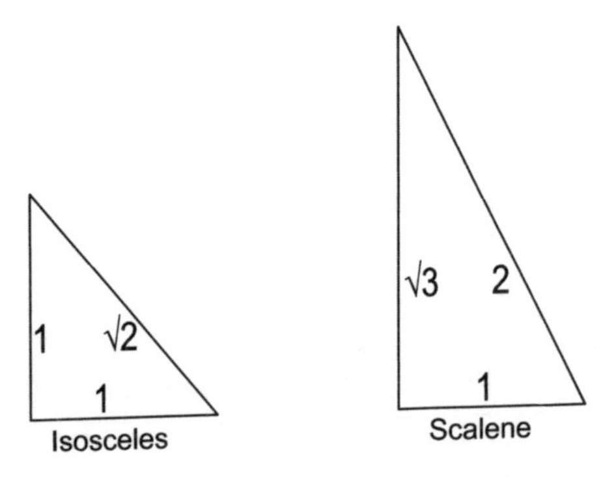

Figure 4: Isosceles and scalene triangles. For Plato, these were the basic entities of the four natural elements.

These two shapes were able to come together to form either squares or a complex triangle (Figure 5).

From these, either a cube, a tetrahedron, an octahedron or an eikosahedron (20-faced object) could be constructed (Figure 6).

Plato associated these four shapes with the standard four elements of the Greeks. The cubes were earth, the tetrahedra fire, the octahedra water and the eikosahedra air. The properties of the elements could be explained by the shape of their particles. Fire could change and cut up other objects, since its particles were small and had sharp corners, whereas earth, as a cube, was solid and could be well packed.

These solids – the cube, tetrahedron, octahedron and eikosahedron – are the perfect, or Platonic, solids. They are made up of faces which are all the same shape and size. A *demiourgos* forming the best possible sort of

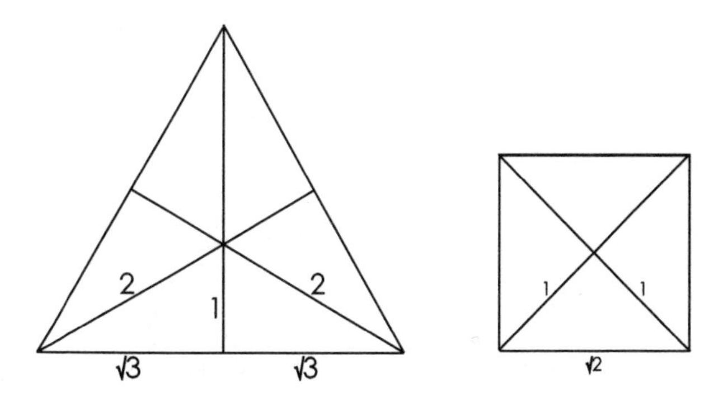

Figure 5: A complex triangle and a square, made up of scalene and isosceles triangles respectively.

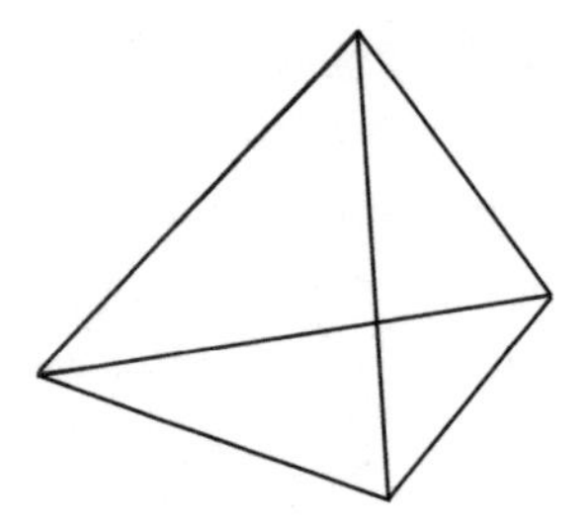
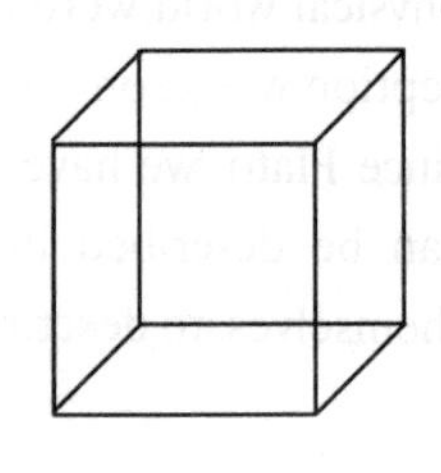

Figure 6: A tetrahedron and a cube – for Plato, these were the basic particles of fire and earth respectively.

world would create these basic building blocks to ensure that the elements were properly proportioned as perfect solids. When he imposed shape and number on the chaos, this is how he did it. He also created a certain order for the earth and for the heavens.

Whatever the merits of Plato's view on atomism, the basic principle that he introduced to science has been important ever since: although the world appears to be diverse and complex, there is an underlying order which we can capture with mathematics and geometry, and that order is simple, elegant and aesthetically pleasing. Scientists today, especially in the more theoretical and mathematical disciplines, still strive for these qualities in their work. Another interesting aspect of Plato's atomism was this. The Pythagoreans, who believed that the world was in some way made up of numbers, were perplexed to find irrational numbers, numbers which cannot be expressed as the ratio of two integers. Plato, by contrast, thought that the ultimate constituents of the physical world were not numbers but shapes. His conception was geometrical rather than arithmetical. Ever since Plato, we have thought in terms of shapes which can be described by numbers, rather than numbers themselves, to describe the ultimate pieces of the world.

No Slight or Trivial Influence

Plato was also keen on the use of geometry and astronomy in education. His training for philosophers led from astronomy to geometry to a contemplation of the nature of the good. He was enormously influential in encouraging the study of these subjects, and was always keen to take them to the next level of abstraction, to establish more general and refined theories. Legend has it that the words 'let no one ignorant of geometry enter here' were inscribed over the door of Plato's Academy.

One can take Plato's philosophy in many directions, and there have been numerous interpreters of his views. There are elements of his thought that can be interpreted in a mystical and religious manner. Early neo-Platonists laid stress on the soul and God, and how we can best come to know God. Their thinking was enormously important for early Christianity, and played down the importance of the investigation of the physical world. Plato himself certainly emphasised the abstract and the theoretical, and so has been seen by some as downplaying the empirical role in science. His positive legacy, though, is the idea of a mathematically structured cosmos, and the notion that mathematics can play a key role in science. These features of his thought were stressed by Renaissance neo-Platonism, and contributed to the emergence of the scientific revolution. They were also an important alternative to Aristotelian ideas on these matters.

Aristotle: The Master of Those who Know

Aristotle was born in Stagira, the son of a doctor to the court of King Amyntas of Macedonia. He was a pupil of Plato at the Academy, where he arrived at the age of seventeen; he studied and worked there for twenty years, to 347 BC. Then came a period of travelling, during which he undertook extensive research in biology, before returning to Athens in 335 BC. He also acted as tutor to Alexander, son of Philip of Macedon, the future Alexander the Great. Aristotle's Macedonian connections may explain why he left Athens – anti-Macedonian sentiment was very strong there at times. Aristotle finally left Athens in 323 BC, and died in 322. Despite being the best student of the Academy, Aristotle was never offered the headship. He founded his own school, the Lyceum, and had a dedicated band of followers throughout the age of antiquity. The sites of the Academy and Lyceum can still be found in Athens.

We have many, but not all, of Aristotle's works. The

style of these is in sharp contrast to that of Plato, but we may well have only Aristotle's lecture notes. Many people in antiquity praised the literary virtues of his finished works, which are now lost. Aristotle wrote on a huge range of subjects, those of particular interest to us being physics, cosmology, meteorology, biology and metaphysics.

At first a follower of Plato, Aristotle later developed his own distinctive philosophy. While he disagreed with Plato on many matters, there were also considerable similarities. Like Plato, he believed that the cosmos, and living things, had not come about accidentally, but required some sort of teleological explanation. Aristotle's views had an unparalleled influence. He established a system for thinking about the world that was dominant for 2,000 years, lasting until the seventeenth century. Above all, Aristotle was a great systematiser. He had great breadth of thought, and also an internal coherence and consistency that made his thinking both plausible and difficult to replace. It could be supplanted only by something at least as comprehensive. Aristotle was the founder of several disciplines, most notably biology and logic. To his later supporters, he was known simply as 'the philosopher' or 'the master of those who know'.

The Terrestrial Realm

Aristotle had to answer the question of why things fall to

the ground, and why the heavens move as they do, without having a theory of gravity. His answer was that everything had a natural place, and a natural motion in relation to that natural place. Aristotle accepted the usual four Greek elements of earth, water, air and fire. The natural place of earth was at the centre of the cosmos, and the natural motion of pieces of earth was in a straight line towards that centre. So any piece of earth not at the centre would naturally move towards it, unless prevented from doing so. Water also had its natural place at the centre, and its natural motion was to move towards it, but it was not as heavy as earth, so it tended to settle on top of it. Objects did not fall because of gravitational attraction, according to Aristotle, but because of their natural place and their natural motion. This piece of ancient physics demanded that the earth be at the centre of the cosmos. If all of the pieces of earth have their natural place at the centre, and a natural motion towards it, then the earth must be at the centre. If it were ever displaced from there, it would have a natural motion to move back there. So Aristotle's physics was inherently geocentric. The earth *had* to be at the centre of the cosmos.

Air and fire were thought by Aristotle to be light – not less heavy than earth and water, as we would say, but positively light. Their natural motion was away from the centre of the cosmos, and their natural place was at the edge of the terrestrial realm. This reached up to the

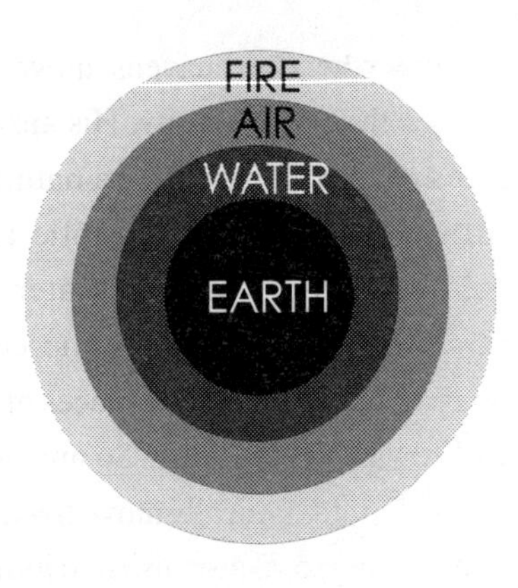

Figure 7: Aristotle's terrestrial realm, as it would be if there were no mixing of the elements.

moon, where the celestial realm of the heavens began. Fire was lighter than air, so it formed the outermost part of the terrestrial realm, with air beneath it. If the elements were to perform only their natural motions, then they would separate out into four concentric rings, with no mixing of the elements.

Aristotle distinguished between 'natural' and 'enforced' motion. If earth was thrown upwards, or sideways, that was contrary to its natural motion. Gradually, natural motion would take over again and it would begin to move towards the centre of the cosmos. Enforced motion required the application of force, while natural motion did not. One point on which Aristotle differed radically from modern views was that he believed that no force was necessary to begin natural motion (or end it when a body reached its natural place). We know that any change in motion requires a force. The elements did not completely settle out into separate bands, according to Aristotle, because of the effects of the sun. The sun had daily and seasonal variations, and its heat stirred up and mixed the elements. The sun, as the cause of daily heating and the seasons, was ultimately responsible for all of the mixing of elements in the terrestrial realm.

The Heavens

So far we have spoken of the terrestrial realm. For Aristotle, the moon, and everything beyond the moon, was the celestial realm. This contained none of the four elements of the terrestrial realm, but was composed of a fifth element known as 'aether'. The Greeks saw the heavens as unchanging. For this belief they had the evidence of their own observations, and those of the Babylonians and Egyptians, of which both Plato and Aristotle were aware. No change in the relative position of the stars had been seen. Aristotle saw the terrestrial realm as a place of change, and change typically took place between pairs of qualities, such as hot and cold or wet and dry. He also believed that no enforced motion – and no natural motion in a straight line – could be permanent. So the unchanging heavens must be made of some other substance, not subject to change, without qualities, and having a natural circular motion. This was aether. Aether was neither light nor heavy (not having the same sort of characteristics as the terrestrial elements), so there was no question of the heavenly bodies plunging to earth. Aristotle also believed there to be a 'prime mover', something which initiated motion in the heavens but was not itself moved (an 'unmoved mover'). Effectively this was god, and god spent all of his time thinking about thinking – the supremely pleasant activity, according to Aristotle. A philosopher's god, if ever there was one! Since Aristotle believed the heavens to be unchanging, he accounted for novae (new stars), comets, etc., by saying that all such phenomena took place in the upper reaches of the terrestrial realm, rather than in the heavens (and these phenomena were likely to be fiery, because that was where fire had its natural place).

Speculations Upon Matter

Aristotle disagreed with Plato and the atomists about the elements. Aristotle was not an atomist. He did not believe that matter came in small, discrete packets, nor did he consider this a useful way to think about the question. He did believe in the usual four elements of the Greeks, but thought of them in qualitative terms. The four essential qualities for Aristotle were hot and cold, and wet and dry. Each of the elements possessed a pair of the contrary qualities. This can be represented schematically (see Figure 8).

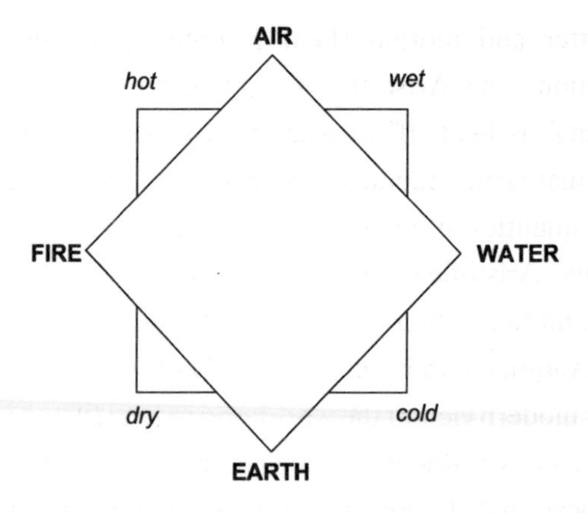

Figure 8: Aristotle's elements. Each element is characterised by a pair of the wet/dry and hot/cold opposites.

These elements could quite readily transmute into one another. By heating water (cold and wet) one could produce air (hot and wet). As well as not believing in atoms, Aristotle denied that there was any such thing as empty space. For him, the world was full (the technical term is a 'plenum'), and a true vacuum did not exist.

Aristotle and Qualities

Aristotle's view of matter was essentially a qualitative one. The world was made up of real qualities. The basic qualities of hot, cold, wet and dry could not be broken down or analysed any further. Where we would say that heat is the agitation of small particles, Aristotle would claim it to be a *quality* of a body. We reduce heat to matter and motion (heat is nothing but matter in motion), but Aristotle thought that there was such a 'thing' as heat. If we were to ask Aristotle what the ultimate constituents of the world are, he would reply: the qualities of hot, cold, wet and dry – and not atoms. Thus, Aristotle's view of the world is said to be a 'qualitative' rather than a 'quantitative' one.

Another important contrast between Aristotle and our modern view of the world, and indeed Plato's view, is the role of mathematics and geometry. Aristotle did not believe that these could provide a good description of objects in the terrestrial realm. He believed that the abstractions of mathematics did not apply to real-life situations. Consider a perfect sphere on a perfectly flat plane, as one might in geometry. There is only one point on this plane where the sphere touches it. But, says Aristotle, in nature we have neither perfect spheres nor perfectly flat planes, and any actual sphere on a flat surface will in fact touch it at many points. In these situations, mathematics describes the ideal and not the actual, and the actual may be very different from the ideal. Aristotle also believed that there were qualities which could not be quantified in any mathematical manner. One might contrast hot, hotter and hottest – that is, make qualitative comparisons – but Aristotle did not think that one could attach numbers to these qualities. One issue here is antiquity's lack of technology for making such quantifications (i.e., thermometers for measuring heat), but at root Aristotle had a different view of what makes up the world. Atoms might be treatable mathematically, but qualities were not – and qualities, for Aristotle, were the basic constituents of the world.

Here is a summary of Aristotle's theory of the elements, including weight, natural place, natural motion and qualities. This theory of the elements and of natural place was retained up to the beginning of the seventeenth century, and the idea of fire as a substance lasted well into the nineteenth century.

Element	**Weight**	**Natural place**	**Natural motion**	**Qualities**
Earth (solids)	**Heavy**	**Centre of universe**	**Straight, down**	**Cold and dry**
Water (liquids)	**Heavy**	**Centre of universe**	**Straight, down**	**Cold and wet**
Air (gases)	**Light**	**Edge of terrestrial realm**	**Straight, up**	**Wet and hot**
Fire (fire)	**Light**	**Edge of terrestrial realm**	**Straight, up**	**Dry and hot**
Aether (stars, sun, moon, planets)	**None**	**Celestial realm**	**Circular**	**None (aether has no qualities)**

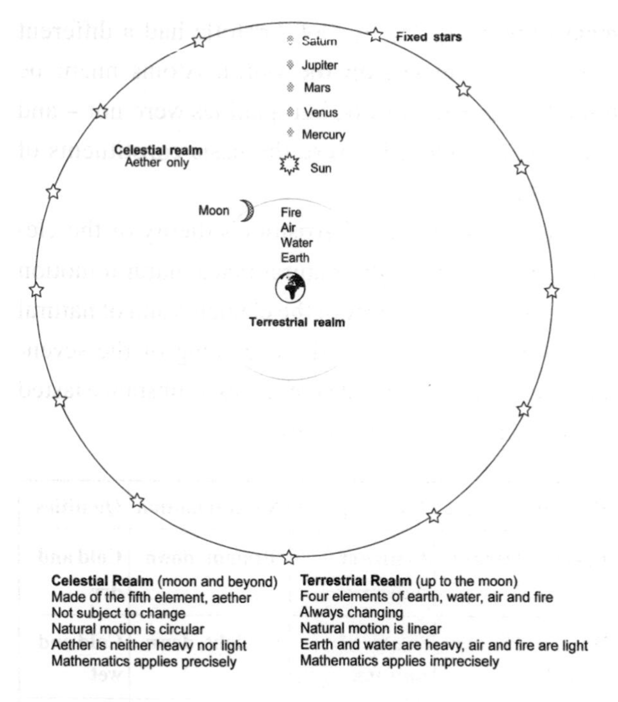

Figure 9: Aristotle's cosmos (not to scale).

Explanations

Aristotle also had a distinctive scheme for explaining the natural world, which was hugely influential. He required four types of explanation for the full description of a phenomenon. These are sometimes known (rather inaccurately) as the 'four causes'. Actually, only one of these explanations is like a 'cause' in the modern sense, and they are better referred to as the 'four becauses'.

Efficient explanation was close to, but not identical with, our notion of a physical cause. *Material explanation* told you what something was made of. *Formal explanation* told you what shape or form something had. *Teleological explanation* told you the purpose that an object or process had. If we think of a table, we can say that it is made of wood (material explanation), it has the form of a table (formal explanation), it was made by a craftsman (efficient explanation) and it is for writing on or eating off (teleological explanation).

Aristotle's teleology differed from that of Plato. While Plato saw teleology as imposed on nature by the *demiourgos*, for Aristotle it was inherent in nature. Nature did not deliberate – that is, there was no conscious design or conscious designer – but many things (in particular, animals) could be seen to have the best possible form. Nature inherently produced what was best of its own nature. This difference in teleology is reflected in the two philosophers' views on the origins of life and the cosmos. Plato believed the cosmos, and human beings, to have been ordered for the best, out of a chaos, by the *demiourgos*. Aristotle believed that both the cosmos and human beings had *always* existed.

There was another important aspect of explanation for Aristotle, and that was the 'potential' and the 'actual'. Everything had a potential to be something or somewhere else, which it might actualise. An acorn, which is potentially an oak tree, would grow into one unless hindered from doing so, and a foal would become a horse. This looks quite handy for biology, but Aristotle also used it for physics. So earth had a potential to be at the centre of the universe, and it fell unless prevented from doing so. All objects had a natural place, and would undergo natural motion to that place unless stopped. The heavens were more actual than the terrestrial realm (where there may be unnatural motion), since they were always actualising their potential. The prime mover was entirely actual, since it had no motion.

Direction of Explanation: Clockwork Lives

This reveals a very important aspect of ancient, as opposed to modern, explanation. Let us ask a deceptively simple question. What is the physical world *like*? Is it like something mechanical (so that it works like a clock) or something organic (so that it works like an animal)? The scientific revolution of the seventeenth century very firmly opted for clockwork. The world was a giant mechanism, and even animals were to be thought of as very complex pieces of clockwork. The modern view is, of course, more sophisticated, but we retain the notion that ultimately the world is made up of inert pieces of matter (atoms, sub-atomic particles) which interact in a mechanical manner. Thus, we might explain a living thing in terms of its component parts, reducing qualities to matter and motion. Analysing a tree, we might move from botany to biology to biochemistry to chemistry, and ultimately to physics. We are happy when we can explain things (at least in principle) in terms of the physics of the ultimate particles.

Not all of the ancients shared this view either of the world or of mechanisms. The sorts of machines that they were acquainted with (wooden carts, etc.) were not paradigms of regularity and precision in the way that clocks are. Regularity, precision and order for the Greeks were the signs of intelligence, not mechanism. The ancients also struggled to explain the growth and behaviour of animals and plants in mechanical terms, and they tried to explain the origins of life and the cosmos as well. Aristotle and many other ancients rejected atomism.

So instead of mechanical explanations being used in biology, they tended to use organic explanations in physics. Where modern explanations tend to be analytical and reductive (we try to explain in terms of the ultimate component parts), many ancient explanations tended to be more holistic, especially among those who rejected atomism. In the case of Aristotle, we see some of the consequences of the lack of a proper conception of gravity and force. He needed the idea of natural place and natural motion, and the idea of the actualisation of potential – both rather organic ideas – to account for phenomena that we would explain by the force of gravity.

Aristotle's legacy was an immensely broad, coherent

and powerful system of the world. His views were those which the scientific revolution had to replace. In explanation, holism was replaced with reductionism, qualities with quantities, organic metaphors with mechanical ones – especially the favourite seventeenth-century metaphor of clockwork. Teleology was rejected, and mathematics, in the form of mathematically framed laws of nature which applied precisely and universally, found a new importance. Atoms with mathematically quantifiable properties replaced qualities, and heliocentrism replaced geocentrism. The distinction between the celestial and terrestrial realms was abandoned, since gravity could now explain both the motions of the planets and the falling of objects dropped on earth.

4 Heavenly Thoughts

But did thee feel the earth move?

Ernest Hemingway, *For Whom the Bell Tolls* (1940), chapter 13

We have already seen something of the beginnings of Greek cosmology, and the crucial change from myth to theory. The pre-Socratics overcame some significant conceptual hurdles to achieve a more sophisticated cosmology. There was the move from a hemispherical universe to a spherical one, and from an earth supported by water to one supported by air, and then to one which required no support. In the earlier Greek cosmologies, objects were thought to drop in parallel straight lines from the top of the cosmos to the bottom. This led to the problem of why the earth, which would seem to be heavy, does not fall to the bottom of the cosmos. In this sort of cosmology, something is required to support the earth. A different way of accounting for the effects of gravity was to have a 'centrifocal' theory. Aristotle placed the earth at the centre of the cosmos, and had heavy objects move towards it. There was now no question of the earth dropping, since, as a heavy object, it moved to the centre of the cosmos, which was itself.

The idea that the earth was central and stable dominated Greek astronomy and cosmology. The Greeks had some good reasons for their belief. We suppose the earth to have two main motions. It spins on its axis once a day and orbits the sun once a year. The Greeks were worried that if the earth was in motion, then there ought to be perceptible consequences. Their experience told them that if you were in rapid motion (and 'rapid' for the Greeks would be horse-riding or running), you certainly knew about it. So they asked: If the earth has a daily rotation (from west to east), why is there not a constant wind (east to west)? If the earth is in motion around the sun, why are objects such as ourselves not swept off the face of the earth? Today we have answers to these problems. We believe that the earth carries its atmosphere with it, and that space is a vacuum, so we see no problem in having the earth (with its atmosphere) spinning on its axis and orbiting the sun. Greek physics had no such answers. The Greeks did not believe space to be a vacuum, but to be full of matter, and so did not distinguish between the earth's atmosphere and space. They did not observe the consequences that they thought should come from the earth being in motion, so they did not believe the earth to be in motion.

There were further problems which occurred to some of the Greeks, which would not occur to someone with a knowledge of gravity. They worried that if the earth was in rapid motion, why did it not disintegrate? For many of the Greeks after Aristotle, the reason why the earth held together, and why objects fell to its surface, was that pieces of earth had a natural motion towards the centre of the cosmos. Move the earth from the centre of the cosmos, and there was no longer any reason why it should hold together, or why objects should fall to its surface. And if the earth was in motion, why did the moon follow it around?

And there was something else. The earth takes up different positions around the sun during the year, and so has different positions relative to the stars. Observations taken six months apart (to maximise the difference) should reveal slight changes in the apparent positions of the stars from earth. This effect is called 'stellar parallax'. However, the Greeks could observe no such parallax effects. This is no great surprise, since we know these effects to be very small, due to the distance of the stars. They were not detected until 1838, by an astronomer called Bessel. Without telescopes, and indeed very sophisticated and powerful telescopes at that, the Greeks had no hope of detecting stellar parallax. Those who believed the earth to be in motion, from Copernicus in 1543 onwards, said that the stars were too far away for stellar parallax to be detected by current means. However, the Greeks believed the cosmos to be relatively small. They believed that the stars were all equidistant, and were beyond the further-most of the planets visible to the naked eye, Saturn. The whole cosmos, for the Greeks, was no bigger than our solar system. They believed that stellar parallax should have been observable, if the earth orbited the sun. It was not, so they believed the earth to be immobile.

The Greeks, then, had several good reasons for believing the earth to be central and stable. Their physics, astronomy, philosophy and common sense all seemed to indicate an immobile earth. What is more, their astronomy seemed to be making great strides forward. There was no reason to suppose that the earth was in motion.

There was an important consequence from this. All of the motions of the heavens were real motions to the Greeks, not apparent ones due to the motion of the earth. If one looks at the stars over the course of a night, they appear to move in a circle. We know that this is due to the fact that the earth spins on its axis. It is the earth that is moving, not the stars. The motion of the stars is apparent, not real. The Greeks, with their faith in the immobility of the earth, believed that it was the stars that were moving in great circles, not the earth. For them, the motion of the stars was real, and not apparent.

At the outset, Greek observational astronomy was rather divorced from philosophical speculation about the nature of the cosmos. There were those who observed the heavens and took careful note of what they saw;

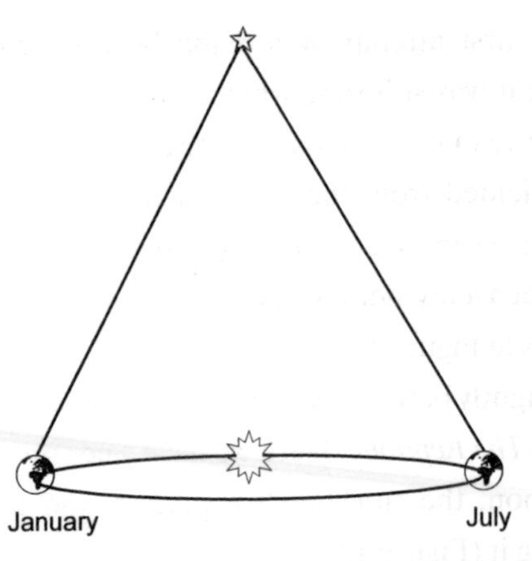

Figure 10: The problem of parallax (not to scale – the star would be very, very much further away!). If the earth orbits the sun in one year, then there is a significant difference in its positions at six-monthly intervals. This ought to be detectable relative to the fixed stars, which should appear to be in slightly different positions. In fact, this is difficult, since the amount that the earth moves is very small compared to the distance of the stars, and parallax was not detected until 1838. Parallax for alpha centauri (our nearest star) = 0.75 of 1" (one second of arc), where 60" = 1' (one minute of arc) and 60' = 1° (one degree). One needs to look at a very near star (4 light years away, not 400 or 4,000), or this effect is not seen at all. The Greeks believed all of the stars to be equidistant and relatively near, just beyond Saturn.

and there were those who produced cosmological models based on general philosophical considerations. No one produced cosmological models that were anywhere near explaining, in a precise manner, the phenomena that had been recorded. This is no great surprise, since the phenomena are quite complex.

The first attempt was made by the Pythagoreans, though it was still somewhat vague and speculative. In the centre of the cosmos was a fire (not the sun), but this was shielded from the earth by a body known as the counter-earth. We never saw this central fire, but the other heavenly bodies revolved around it, outside the earth (see Figure 11).

A slightly better model of the heavens can be found in Plato's *The Republic*. Here we have a central earth, with the moon, the sun, the five planets and the stars all orbiting it (Figure 12).

You will notice that in both of these models, all of the motions of the heavenly bodies are assumed to be

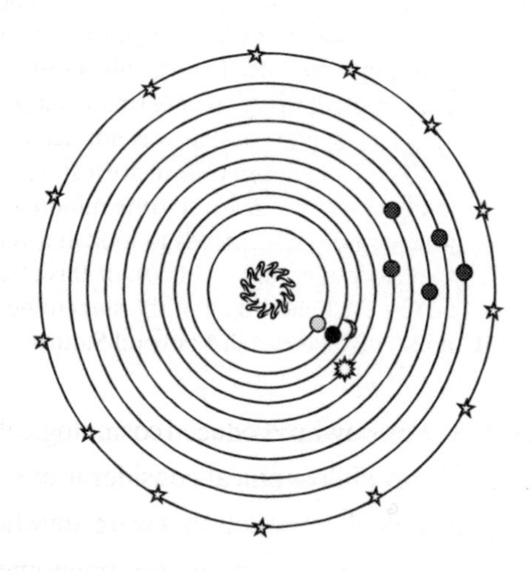

Figure 11: The Pythagorean cosmos, showing the central fire, then the counter-earth, earth, moon, sun, five planets and stars.

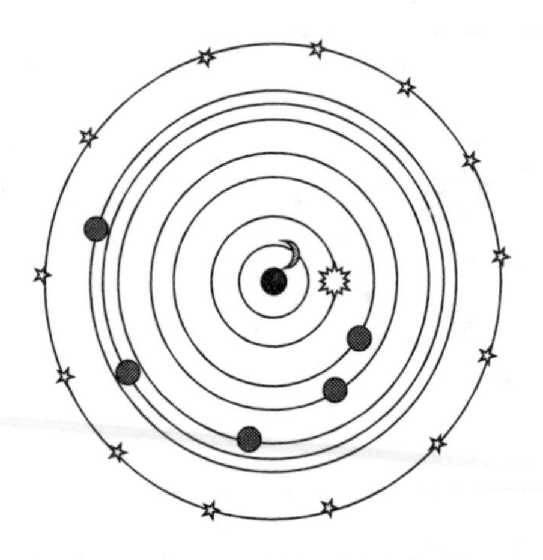

Figure 12: Plato's early view from *The Republic*. A central earth, then the moon, sun, five planets and stars.

regular and circular. This was a basic assumption of Greek astronomy and cosmology. Why? Simply because the Greeks considered this to be the best sort of motion. Circular motion could continue without change, and it had a simplicity and elegance which appealed to them. A well-ordered cosmos, such as the one that the Greeks believed themselves to live in, would see the heavens moving with regular circular motion.

Neither of these two models could account for two important phenomena relating to the point at which the sun sets. The sun does not always set due west. If one takes note of where on the horizon the sun sets during a year, this changes from a maximum of 23.5° north of west to a maximum of 23.5° south of west. Solstices (shortest

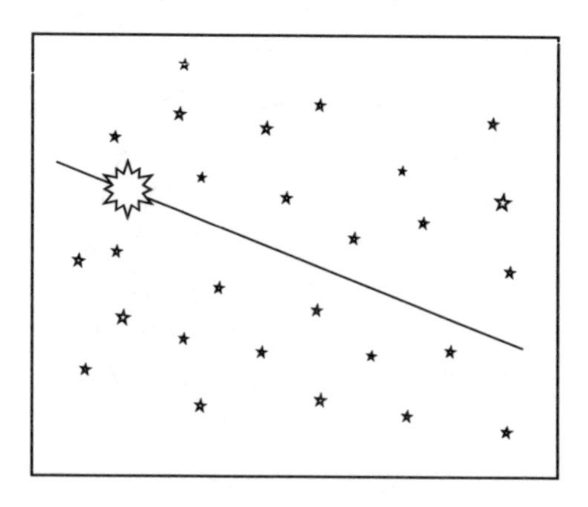

Figure 13: The sun's motion against the background of the fixed stars, tracing out a line called the 'ecliptic'. One observes where the sun sets and which stars then appear at that point.

day or night) occur at the maximum points, while equinoxes (equal day and night) occur when the sun sets due west. If one takes note of which stars first rise at the point on the horizon where the sun sets, these can be seen to change during the year as well. These phenomena were well known to the Greeks and many other ancient societies.

The first model that could give a reasonable account of these phenomena, and which was perhaps the first serious attempt to unite the astronomical and cosmological traditions, came with Plato's book *The Timaeus*. Here he introduced some very important ideas. The stars seemed to the Greeks to show good order. They moved in what appeared to be perfect circles. However, the five planets that can be seen with the naked eye – Mercury, Venus, Mars, Jupiter and Saturn – all have motions relative to the stars. These motions are quite complex, and initially did not seem orderly to the Greeks. Indeed, our word 'planet' comes from the Greek '*planetes*', which means a wanderer or a vagabond. The Babylonian word for a planet was '*bibbu*', meaning sheep. Plato insisted that the planets did not in fact wander, but moved in orderly but complex combinations of regular circular motions. This set the terms for astronomy for two millennia. Not until 1609, when Kepler recognised that planetary orbits are ellipses around the sun, was this to change.

The essence of Plato's later model was that the sun, moon and planets have a combination of two regular circular motions. The stars are still carried around once a day in one motion, but the sun, moon and planets have a second motion in addition to the daily one, offset at an angle to it. So they had motion relative to the fixed stars, as well as moving with the stars (see Figure 14).

This model gave a reasonable approximation of the setting sun phenomena, but could not explain everything. If you watch the motions of the planets against the background of the fixed stars over a year or two, you will see something strange. Normally, the planets will progress against the background of the fixed stars. However, they will sometimes come to a halt, go in the other

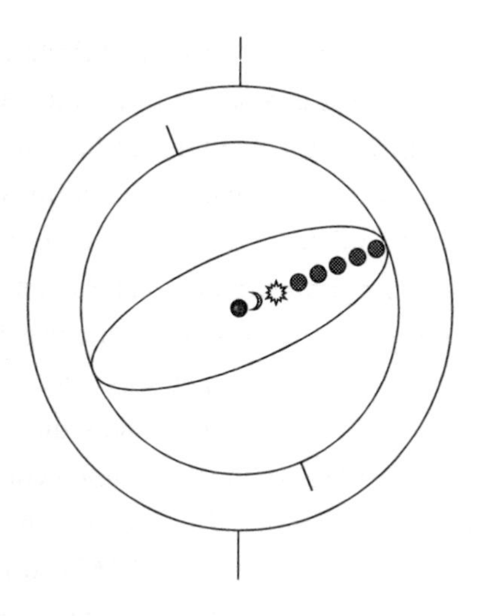

Figure 14: Plato's later model, employing two spheres for moon, sun and planets. The earlier Greeks assumed the angle between these motions to be 24°, 1/15 of a circle.

direction for a while, stop again and then go in their normal direction once more. This is called 'retrogression' or the 'retrograde motion' of the planets (Figure 15).

The planets do not follow the ecliptic (the path of the sun against the background of the fixed stars) exactly. They deviate a few degrees either side of it. The band within which the planets move is called the zodiac. This band can be split into twelve parts to give the houses of the zodiac.

Plato's model, though an advance in astronomy, was still qualitative and did not account for either retrograde

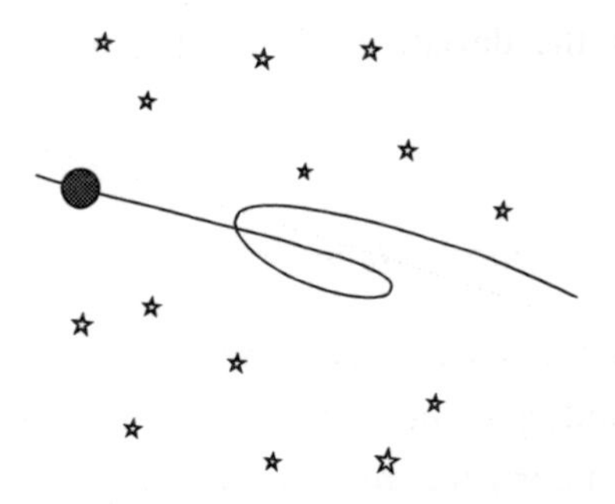

Figure 15: A planet undergoing retrograde motion. Planets do not actually stop and then move on again. They appear to do so to someone on earth, because the earth and planet have different sizes and speeds of orbit around the sun. Sometimes these combine in such a way that the planet appears to come to a halt, reverse its direction, and then move on as normal. Because the Greeks believed in an immobile earth, the planets for them had to have real retrograde motion.

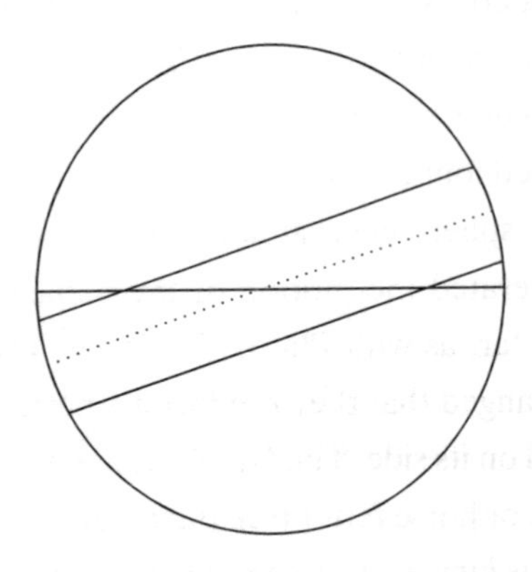

Figure 16: The ecliptic and zodiac are often represented like this. The zodiac can be divided into twelve parts, as is familiar from astrology.

motion or the deviation of planets from the line of the sun.

Eudoxus: Thinking Regressively

While Plato may have been important in formulating the ideas underpinning ancient astronomy, undoubtedly the greatest early theoretician was Eudoxus of Cnidus (*fl.* 365 BC), who was also a brilliant mathematician. Eudoxus was a pupil of Archytas the Pythagorean, and seems to have had a close relationship with Plato. He travelled widely, made astronomical observations and founded a school at Cyzicus. He took Plato's model and made it more complex and much more accurate. The next part is a little tricky, but worth following to get a sense of Eudoxus' genius and the way in which the Greeks went about astronomy. While Plato's model had two regular circular movements for each planet, Eudoxus used four (see Figure 17).

The first sphere generated a daily motion, and the second generated the motion of the planet along the ecliptic (so far, as with Plato). The other two spheres were so arranged that they produced a pattern like the figure 8 laid on its side. The Greeks called this pattern a '*hippopede*', or horse fetter (see Figure 18).

When this *hippopede* is combined with the other two motions, you get a pattern that looks very like regressive motion (Figure 19).

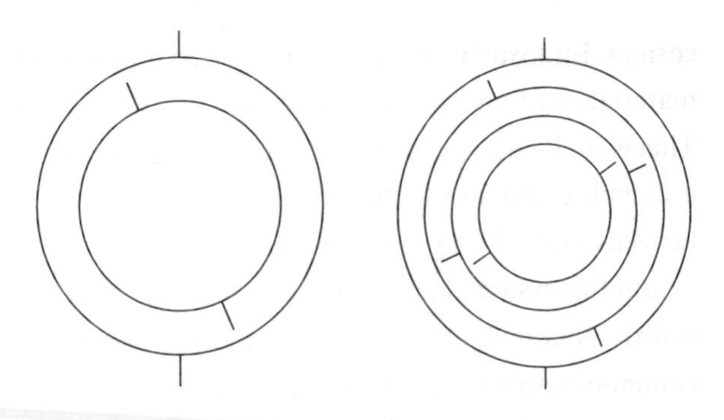

Figure 17: Plato's two-sphere model, and Eudoxus' four-sphere model for a planet.

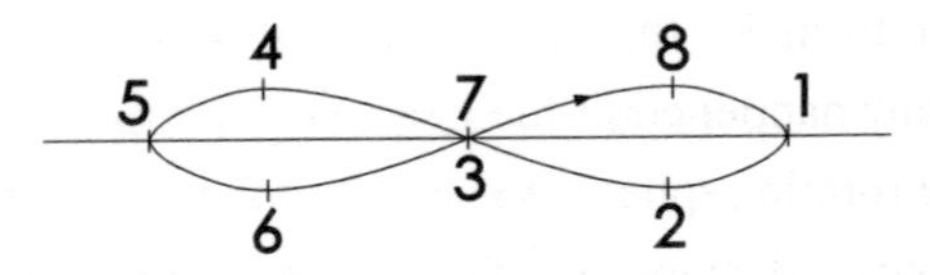

Figure 18: The shape generated by two of Eudoxus' spheres, known as a *hippopede* (horse fetter).

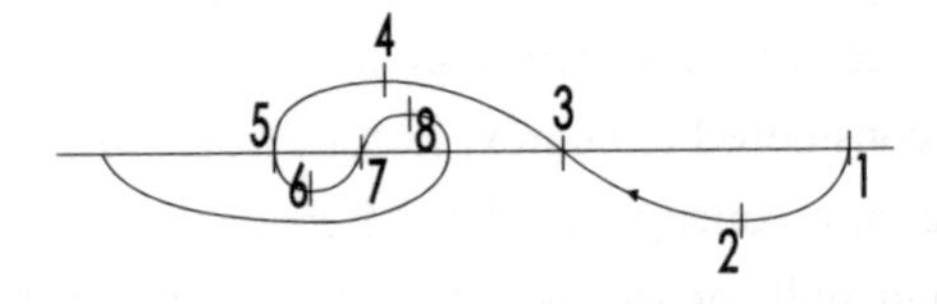

Figure 19: The resultant motion of Eudoxus' four spheres.

This allowed Eudoxus to cope with some of the main deficiencies of Plato's model. In one of the most brilliant pieces of ancient science, Eudoxus did the mathematics required to make this system work as a real model of the cosmos. Eudoxus' resources were some primitive writing materials and a record of the motions of the heavens. That is all. It is quite amazing that he was able to produce a complex and workable mathematical model of the heavens with nothing more than this.

But this system, known as the 'concentric' or 'homocentric' sphere system – since all of the spheres have a common centre, the centre of the cosmos – was far from perfect, and Eudoxus was well aware of this. One of Eudoxus' pupils, Callippus of Cyzicus (*fl.* 330 BC), made further changes to his scheme of rotating spheres in order to make the model fit more closely with what actually happened in the heavens. He introduced even more rotating spheres to account for some quite subtle changes in the motions of the planets. Aristotle made no improvements to concentric sphere astronomy, but thought hard about the cosmology. He conceived of each of the spheres of Eudoxus and Callippus as being real and made out of the fifth element, aether. These spheres were considered to be next to each other with no space between, a conception of the heavens as being like a Russian doll, or the layers of an onion. Each of the spheres then contributed to the motion of its planet.

The two defining features of Greek astronomy were an immobile earth and a belief in circular motion. They treated all of the motions of the heavens as real motions, and as combinations of regular circular motions. Greek astronomy became even more complex as it tried to reproduce the motions of the heavens with ever greater accuracy. The other important feature of ancient astronomy was that it was naked-eye astronomy. The Greeks did have some devices to help them observe the heavens more accurately, but no means of image intensification. The first telescopes were not invented until 1609, leading to the discovery of a great amount of new information simply unknown to the Greeks.

Ptolemaic Astronomy

The system devised by Eudoxus and Callippus was an excellent model for that time, but there were some inherent difficulties. Planetary orbits are in fact elliptical around the sun, and not circular around the earth. This means that planets get nearer and then further away from the earth, such that their apparent size varies. The concentric sphere model has the planets at a constant distance. As a planet goes around its real orbit, its apparent speed varies – faster when it is nearer the sun, slower when further away. The concentric sphere model has great difficulties in coping with this. The planets have different shapes of retrogressive motion, while the *hippopede* can give only one shape.

Building on the work of Apollonius of Perga (262–190 BC) and Hipparchus of Nichaea (*fl.* 135 BC), Ptolemy of Alexandria (*c.* 100–170 AD) produced a whole new system which was to last for nearly 1,500 years. Ptolemy observed the heavens from Alexandria in Egypt. He produced a book called the *Suntaxis* in Greek (meaning 'great collection') and the *Almagest* in Arabic ('the greatest'). It was, quite simply, the peak of ancient astronomy. Ptolemy observed, invented new theories and synthesised his own work with what was already known. He was also an excellent geographer, and produced an atlas of maps of the known world.

Ptolemy's system was still based on combinations of regular circular motions. He gave up the concentric sphere and *hippopede* model in favour of a system based on a device known as the 'epicycle' (Figure 20).

The epicycle is a combination of two regular circular motions, but not around the same centre. The centre of

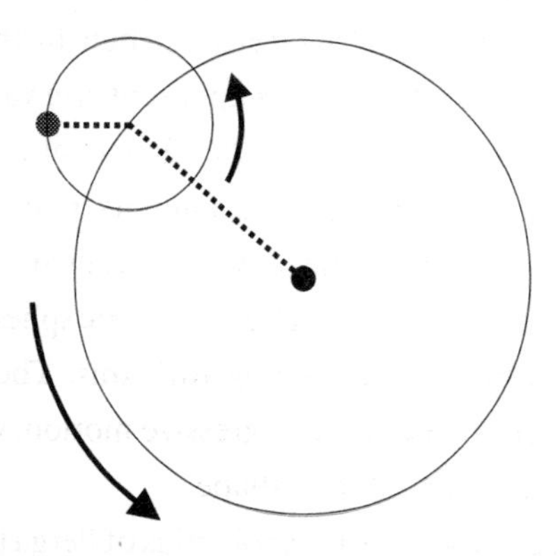

Figure 20: The epicycle, the basic unit of Ptolemaic astronomy. The planet moves around the small circle, whose centre moves around the large circle, giving a combination of regular circular motion.

the smaller circle moves around the larger circle. The actual motion of the planet will depend on the size and speed of rotation of both of the circles, and some quite complex patterns can be produced (see Figure 21).

With this, allied with two more complex devices based on the epicycle, known as the 'eccentric' and the 'equant', Ptolemy was able to account for most of the problems that beset the concentric sphere model. Planets could

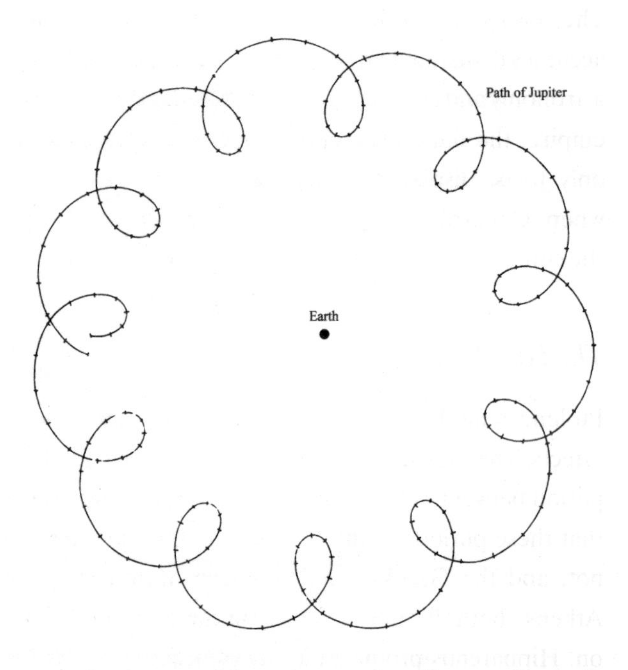

Figure 21: One possible pattern produced by an epicycle. By varying the size and speed of the circles, many others are possible.

them to conceive of a force, emanating from the earth or anywhere else, that would produce these complex motions. They had two sorts of explanation for these phenomena. Plato believed that the planets were intelligent, to the extent that they would always choose the best motion; Aristotle believed that the spheres which moved the planets were made of aether, and the natural motion of aether was in a circle. The result of both explanations was that the planets always moved in an orderly manner with combinations of regular circular motion.

The Greeks are sometimes said to have an 'organic' picture of the cosmos. That is, instead of believing the cosmos to be inanimate and to work in the manner of a machine, they thought of it as more like a living entity. Why did they take this view? To them, the cosmos had certain properties. It was a finite, enclosed, unitary thing. It was well-ordered. The motions of the heavenly bodies were orderly and regular, and the Greeks associated such changes with intelligence. The cosmos appeared to be able to sustain itself in this well-ordered state. So if you had asked the Greeks what the cosmos was like, many would have said that it was like a living thing *in these respects*. This was no primitive anthropomorphism, but an attempt to understand the orderliness of the cosmos in the absence of modern ideas about gravity, force and the laws of nature. Other Greeks suggested that the cosmos was like a political entity, since they were keen to emphasise the orderliness and sophistication of the city in contrast to the chaos and crudeness of the countryside. Others considered the cosmos to be an artefact, something showing the marks of a craftsmanlike creator. If you find these ideas odd, remember that with 'big bang' cosmology, we believe ourselves to live in the aftermath of an explosion. What the universe is *like* is a very tricky question to answer.

What was the picture of the cosmos at the end of antiquity? It was largely Aristotelian. The earth was at the centre, surrounded by the moon, sun, planets and stars moving in a circular manner.

Astronomy was slightly more complex, since Ptolemy's epicycles had proved more fruitful than the concentric sphere system. The size of the epicycles was used

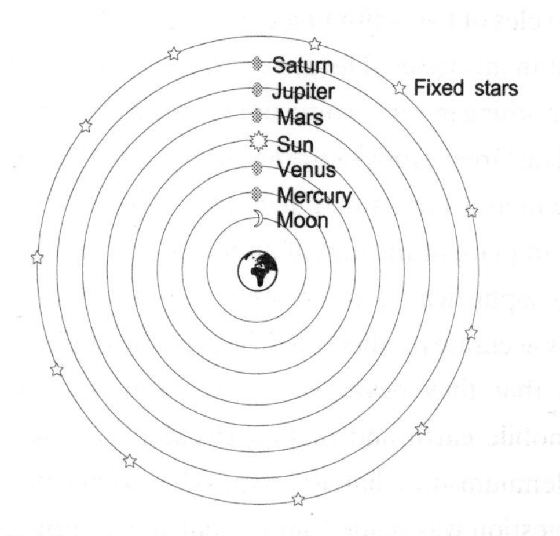

Figure 24: The picture of the cosmos at the end of antiquity. Note that the order of the planets is slightly different from that of Aristotle, who had the sun directly after the moon.

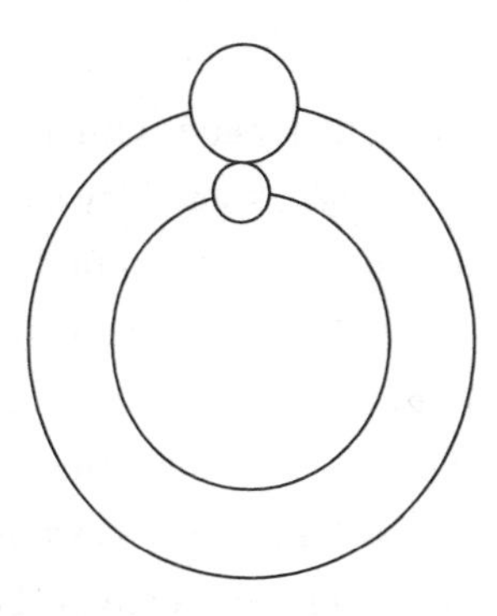

Figure 25: How the ancients spaced orbits. The epicycles touch, but do not overlap, thus giving the spacing. This is very much simplified, but shows the general principle. The key philosophical idea behind this was that there should be no empty, disused space.

to produce a spacing for the planets. The radii of the epicycles of two adjoining planets gave the gap between their main circles. The principle here was that god, who did nothing in vain, would not create empty space.

The Greek study of the heavens was certainly remarkable in its accuracy and ingenuity. The system that they had in place at the end of antiquity was mathematically very sophisticated and powerful, and capable of extremely accurate predictions. The great problem, of course, was that they never got away from the idea of an immobile earth and regular circular motion. Another millennium and a half was to pass before the first serious suggestion was made that the earth was mobile, and it took another seventy years for the problems with this idea to be solved, and for it to become accepted.

5 The Origins of the Cosmos and of Life: Consider your Origins

Except the blind forces of Nature, nothing moves in this world which is not Greek in its origin.

Sir Henry Maine, *Village Communities*, 3rd edition (1876), p. 238

A question which divided the ancient Greeks was how the cosmos acquired its order, and how it was maintained. They effectively split into two camps on this issue. There were those who believed that something directed the cosmos such that there was a good, well-ordered result. Most influential here were Plato and Aristotle. In the opposing camp were those who believed that order came about by chance, the key thinkers here being the atomists Leucippus and Democritus. Along with this question of order was the question of whether there was one cosmos or many '*cosmoi*'. Plato and Aristotle firmly believed that there was one unique cosmos that was in some way structured for the best. The atomists, on the other hand, believed that there were many *cosmoi*, separate from one another, in which everything happened by chance. Their view was that:

> *There are innumerable cosmoi, which differ in size. In some of these there is no sun or moon, in some they are larger than ours and in some more numerous. The spaces between cosmoi are not equal, in places there are more and in others less, some are growing, some are in their prime, some declining, some are coming to be and others failing. They are destroyed by falling into each other. There are cosmoi bereft of animals and vegetation and all moisture.*

The atomists' explanation of the order of our cosmos was that among the infinite number of different *cosmoi*, there would be all of the possible arrangements of a cosmos, including the one that we live in and call a 'good' arrangement.

To see how the atomists' view works, we need to look at what they thought about the origins of the cosmos. Plato, as we have seen, believed that his craftsman god had ordered the cosmos out of a primordial chaos, while Aristotle believed that the cosmos had always been ordered as it is now, and would always be so ordered. The atomists Leucippus and Democritus believed that matter formed a vortex, in which it was whirled around. In this vortex, matter was sorted according to the like-to-like principle. Some types of atoms linked together to form a membrane around a cosmos, and when that happened a new cosmos was formed. The atomist account of such linking was crude. Since atoms had all sorts of shapes and sizes, some had hooks and some had eyes, and so they linked together. Gradually, *cosmoi* separated out from the vortex. There was no limit to this process, and it happened continually, so there were an infinite number of *cosmoi*. Much here depended on the precise effect of the like-to-like principle. A later philosopher, Sextus Empiricus (*fl.* 200 AD), tells us that:

> *Democritus has confirmed this opinion, and Plato has mentioned it in his* Timaeus. *Democritus bases his argument on both living and inanimate things. Animals, he says, gather together with others of like species, as doves with doves and cranes with cranes, and so with other irrational animals. It is the same with inanimate things, as can be seen in the case of seeds which are being winnowed and pebbles on the sea-shore. The whirling of the sieve separates lentils with lentils, barley with barley and wheat with wheat, and due to the motion of the waves, oblong pebbles are moved into the same place as other oblong pebbles, and round with round, as though the similarity possessed by things leads to them being gathered together.*

According to the atomists, the actions of this principle produced our world, as one possibility among many others. As long as there were enough *cosmoi*, chance and the like-to-like principle would bring our cosmos into being. Plato agreed that this principle operated, but disagreed on what the result of its actions would be. He emphasised that the cosmos consisted of unlike things brought together in due harmony and proportion, and did not see how a like-to-like principle could achieve that.

A question that one might ask here is this: Did no one believe that there was just one cosmos that did not require any teleological ordering? In fact, this view is very rare in the history of science. No one in the ancient world believed it. Virtually no one believed it in the era when cosmogony and cosmology were dominated by Christian thought, since the cosmos, in this system of thought, came ready formed by a good God in the six days of creation, as indeed did human beings in the shape of Adam and Eve.

Some alternative early Greek views were those of Empedocles and Anaxagoras. Empedocles believed that there were the four elements of earth, water, air and fire, and that their ordering was governed by love and strife. Love brought things together and strife drew them apart again, and the cosmos underwent a continuous cycle of the dominance of love giving way to the dominance of strife and then vice versa. Two problems with many cosmogonies were pointed out by Plato and Aristotle. Aristotle questioned why motion began, while Plato questioned why, when motion began, it was assumed to have a specific form, such as the vortex of the atomists, rather than just be chaotic. Anaxagoras of Clazomenae (*c*. 500–428 BC) tried to get around these difficulties by suggesting that a cosmic intelligence initiated motion but then took no further part in the running of the cosmos.

Anaxagoras' position has some interesting similarities with those adopted by certain mechanical philosophers in the seventeenth century. For them, a Christian God created the world in a ready ordered state (in line with the book of Genesis), and then took no further part in the running of the universe. For Anaxagoras, cosmic intelligence ('*nous*') provided the initial impetus and order before withdrawing. It is interesting to note that, like a Christian God, *nous* knows all and is all-powerful. The critical difference is that in the seventeenth-century conception, God also created a framework of physical law and forces to which matter is subject. Given the initial ordering, as in the book of Genesis, these forces and laws were supposed to be sufficient to explain the ongoing order of the world. As in the ancient world, there was a religious debate around this idea. Newton and his supporters argued that removing God from any part in the running of the world would lead to atheism. Leibniz, on the other hand, argued that to suggest that God was an incompetent craftsman who would produce a mechanism that needed 'winding up and cleaning' was derogatory of the power of God, and would also lead to atheism.

The Greeks struggled to give a 'one cosmos with no teleology' account of the universe. Some went in for many *cosmoi*, while others opted for teleology to explain one cosmos. A similar question arose concerning the origins of life. There were those who believed that human beings originated to some extent by chance. They believed that there were many different combinations of the available parts of animals before self-sustaining human beings were produced. There needed to be many of these mutations, just as there needed to be many *cosmoi*, to explain how anything so apparently well-ordered as humans had come about by chance. Plato and Aristotle believed that human beings were the result of teleology. Equally, they believed that there was a sense in which humans – and indeed other animals – were unique. There was not a whole spectrum of close relatives which had not quite worked: teleological ordering had created the best arrangement straight away. Aristotle says this about living things:

> *The lack of chance and the serving of ends are found especially in nature's works. The end for which something has been constructed or has come about belongs to that which is beautiful.*

As an example from the other side, let us look at some fragments of Empedocles on the origins of humans:

> *Empedocles believed that the first generation of animals and plants were not generated complete in*

all parts, but consisted of parts not joined together, the second of parts joined together as in a dream, the third of wholes, while the fourth no longer came from homogeneous substances like earth and water, but by mingling with each other.

This leads to a rather nightmarish scenario:

On the earth there burst forth many faces without necks, arms wandered bare bereft of shoulders, and eyes wandered needing foreheads.

Many sprang up two faced and two breasted, man faced ox progeny, and conversely ox headed man progeny.

Eventually, human beings with the ability to reproduce would be formed, and this process would come to an end. The problem here is that the Greeks had no concept of evolution, or of genetics. So those who believed that human beings came about by chance had to struggle very hard to come up with some sort of plausible account of their origins. If Plato and Aristotle found these accounts implausible, I think we may well have some sympathy with them. Plato, who was an excellent satirist, was savage about Empedocles' view. He developed an account of how the *demiourgos* constructed humans with the best arrangement in mind. First the head is constructed, and then we are told that:

In order that it should not roll around on the ground, with its heights and depths of every kind, and be at a loss in scaling these things and climbing out of them, it was given a body as a means of support and for ease of travel.

Aristotle believed in the fixity of species. The cosmos had always existed, as had man and the animals. There was no evolution, and although nature did not deliberate, humans and animals were organised for the best. The teleological accounts of humans that Plato and Aristotle gave must be seen as at least as plausible as the mechanistic accounts of other ancient thinkers, in the context of the resources available in ancient Greece.

There was another area of contention between Plato and the atomists. Plato believed that there were a small number of well-ordered basic particles out of which the physical world was constructed. As we have seen, he believed that there were two basic types of triangular particle which came together in various ways to produce the elements of earth, water, air and fire. Leucippus and Democritus, by contrast, believed there to be an indefinite number of shapes and sizes for their atoms. Plato explained the forms of his basic particles in overtly teleological terms. God created these particles because they have the best shape.

While the scientific revolution adopted atomism and a mechanical account of the cosmos (rather like that of the ancient atomists), it is important to remember the following point. The scientific revolution occurred in a strongly Christian context, in which God created the cosmos ready formed, life ready formed, and was supposed to be responsible for the shapes and sizes of atoms as well. The problems with ancient atomism were resolved by supposing the existence of a God who orders things for the best, and so the influence of Plato and Aristotle was still felt. It was not until much later that theories of the origins of the cosmos and of life were able to do without such an idea.

Remarkably, there is a similar debate in modern cosmology. The problem is slightly different, but the proposed solutions, in their structure, are rather like those pioneered by the ancient Greeks. We can now explain, at least in outline, the origin of the cosmos using gravity. According to modern cosmology, the universe began with the big bang around 15 billion years ago. After the initial expansion there was a period of rapid inflation, followed by a calmer period of expansion. At this time, there were the fundamental particles and there was a great deal of radiation, but none of the elements had formed as yet. This 'chaos' gradually sorted itself as the universe expanded and cooled. Firstly, hydrogen and helium nuclei formed, and then matter and radiation decoupled, with electrons binding to the nuclei of the light elements to form the first atoms. It is from this period, about 300,000 years after the big bang, that we can find evidence of the cosmic background radiation. While this is in remarkable accord with predictions, there are slight inhomogeneities. These small variations in the intensity of the cosmic background radiation, 'ripples', 'wrinkles in time' or whatever one wishes to call them, are the seeds for the future development of the cosmos. Clusters of matter form from these ripples and eventually create stars and galaxies. Stars burn their fuel of hydrogen and helium by nuclear fusion, creating the even-numbered elements up to and including iron. When these stars become supernovae, the other odd-numbered elements, and those heavier than iron, are created. Our solar system is formed out of the detritus of such stars. When sufficiently complex organic chemicals have formed, life gets under way and evolves, which brings us to the present day.

This solves the ancient problem of how the cosmos acquired its order. There is, however, a different modern problem. We are aware that there are critical constants in nature, known as the 'fundamental constants'. So, for instance, the intensity of gravity per unit of matter is a constant (the gravitational constant, $G = 6.7 \times 10^{-11}$ Nm^2kg^{-2}), and the velocity of light is a constant ($c = 3 \times 10^8 ms^{-1}$). What we do not understand is *why* these fundamental constants have these specific values. Nothing determines that they have them, and they could have a whole range of other values. If these values were slightly different, our universe would be very different indeed.

We know that the universe is expanding, and we also know that the rate of expansion is critically dependent on the value of the gravitational constant and the amount of matter in the universe. If gravity were significantly stronger, the universe would have collapsed long before the conditions for the origins of human life had come about. If gravity were significantly weaker, the universe would expand so rapidly that galaxies, stars and planets would not form.

The carbon that is critical to carbon-based life-forms such as ourselves has not always existed. It was produced by a process known as 'stellar nucleosynthesis' in stars. In fact, all of the elements heavier than hydrogen and helium (the two lightest elements) are produced by stars. The production of carbon is part of a chain. Hydrogen atoms fuse together to form helium, then helium undergoes fusion to form beryllium ($2He^4 > Be^8$). The beryllium is short-lived, though, and very quickly another fusion reaction takes place between beryllium and helium to form carbon ($Be^8 + He^4 > C^{12}$). The energy levels mean that this reaction proceeds quickly and very little beryllium is left. Some, but not all, carbon is eventually fused into oxygen ($C^{12} + He^4 > O^{16}$). This reaction is not so efficient, which means that some carbon is left unburnt. So, some of the elements are favoured in this chain – that is, a large amount of them is produced from the previous member of the chain, but relatively little is converted into the next member.

Fortunately for us, carbon is reasonably well favoured. This is critically dependent on the energy levels in carbon atoms, which in turn are dependent on the strength of gravity and the electrical forces in the carbon atom. So our universe is critically dependent on the values of the fundamental constants.

There is a split in modern cosmology. Some, rather like Plato and Aristotle, believe that we need some extra explanatory principles in order to explain why the one cosmos has these characteristics. Answers vary from there being a God who set the values of the fundamental constants in order that human life could come to exist, to some version of the anthropic principle. The anthropic principle is a sort of modern teleological principle which assumes that the cosmos must be such as to allow the eventual existence of humans. Others, rather like the ancient atomists, believe that our universe is only one among infinitely many others, the others differing in the values of their fundamental constants. The modern term for a collection of universes is a 'multiverse'. There is a third approach to this problem which is distinctly modern rather than ancient. This approach says that we simply do not know enough about the universe yet, and that our physics is incomplete. Only when we have the complete physics, and we have a 'theory of everything' (a 'TOE'), will we understand why the fundamental constants must have these values. It was a weakness of ancient science that each of the Greeks tended to believe

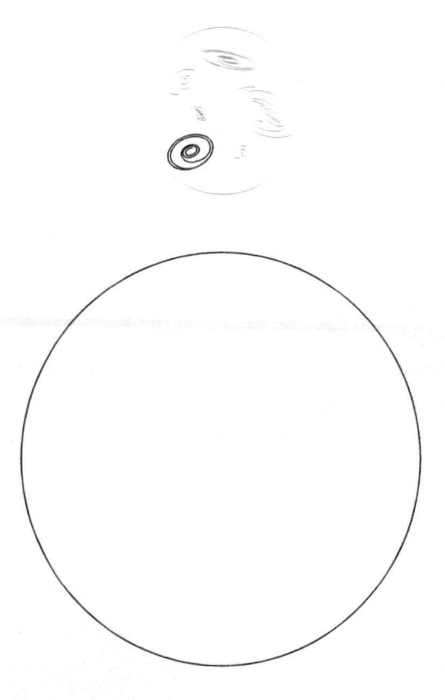

Figure 26: The modern cosmological problem. If the value of the gravitational constant is too great, then the universe will collapse again after the big bang and nothing interesting will be formed. If the gravitational constant is too low, then the universe will expand too rapidly for galaxies, stars and planets to form, and there will be no life. So does the gravitational constant have just the right value, and just the right relation to other fundamental constants, to allow the formation of galaxies, stars and planets – and ultimately life? According to some, we need further principles, such as the anthropic principle, to explain this, or perhaps we need to think of a God deciding on these values. According to others, our universe is just one part of a 'multiverse' of different universes, each with different values of the fundamental constants.

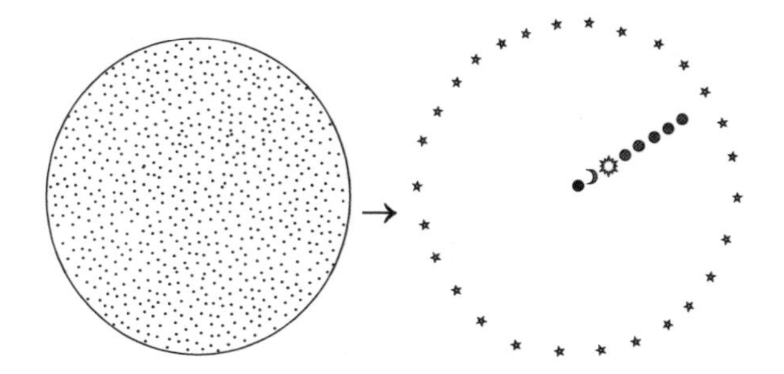

Figure 27: The ancient cosmological problem. Why does chaos organise itself into a well-ordered cosmos (like that of Plato or Aristotle, shown here), rather than simply stay as chaos? Or, if a like-to-like principle is operating, why doesn't the cosmos simply separate out into the constituent elements?

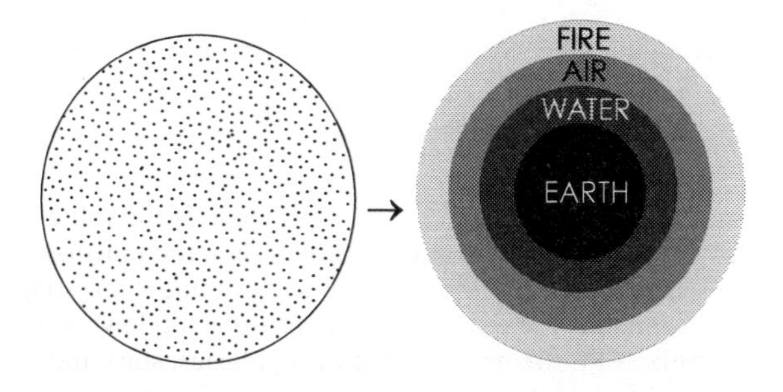

Figure 28: The answer of Plato and Aristotle was that something more was required to explain how the universe was well-ordered. Plato saw good order as imposed by his craftsman god, the *demiourgos*, while Aristotle thought that good order, or the ability to develop towards the best order, was inherent in nature. The answer of the ancient atomists was that there were an indefinite number of *cosmoi*, some chaotic, some with partial order, and so the order of our cosmos had come about by chance.

that they had produced a final and definitive account of the cosmos, rather than contributed something to our knowledge in an ongoing programme of research.

The origins of life are a different matter, since we now have a good account of both origin and evolution, the key mechanisms being genetics and DNA. However, these are relatively recent developments, and even after the scientific revolution, people struggled to give a convincing account of the origins of humans. During the seventeenth century, mechanical biology was developed, functioning on two levels. On a 'macro' level, it treated the body as a machine: limbs and muscles were levers and ropes; the circulation of the blood was a hydraulic system driven by a pump (the heart); and the stomach was a device for crushing and pulverising food. On a 'micro' level, the ultimate goal was to explain the functioning of the body in terms of the matter and motion of its constituent parts. Since matter was conceived to be passive, ultimately all explanation had to be in terms of immediate, mechanical causes.

There were, however, certain very important phenomena that mechanical biology struggled to explain in anything like a plausible manner. Some of these were standard ancient objections, some were generated by the use of the microscope, newly invented by Leeuwenhoek. The basic problem for mechanical biology was that organisms appeared able to organise themselves beyond anything that mechanical biology could explain in terms of matter, motion and mechanical causes. This manifested itself in the question of reproduction, in which things as (mechanically) similar as equine and human embryos in things as (mechanically) similar as equine and human wombs would consistently grow into things as different as horses and humans. A further problem lay with the development of the human embryo, which in its early stages looks nothing like a human being at all, and seems to develop due to some internal dynamic rather than being determined by outside mechanical causes. The ability of some animals to regenerate significant body parts (such as some crabs), or reorganise into two new organisms after being divided (some worms), or reorganise themselves after being turned inside out (water hydra), also indicated a greater ability to self-organise than the mechanical philosophy could allow. A further blow to the mechanical programme was the discovery of the cell, thanks to the new microscopes. The cell appeared to be a living entity in itself, whereas the parts of the body were supposed to be simply mechanical.

So mechanism could not account for the organisational abilities of animals and cells, nor their ability to pass this on. Even Aristotelian ideas of form and the actualisation of potential seemed to provide a better conceptual framework for understanding these phenomena than did mechanical biology. Biology in the eighteenth century therefore adopted vitalist ideas.

The Greeks, and Aristotle and Plato in particular, are often criticised for their teleological approach. Yet it was a popular approach in the ancient world, and indeed for nearly 2,000 years up to the scientific revolution. It is easy to see why. Firstly, purely mechanical accounts were weak in the ancient world. The Greeks simply did not have the resources we now possess to explain the order that we see about us. Lacking a conception of gravity, they struggled to explain how the cosmos could have acquired its current order. Lacking any conception of genetics and DNA, or any serious idea about evolution, they struggled to explain how human beings might have originated. So we see a split between those who believed that these things came about by chance and postulated multiple *cosmoi*, and those who believed in a teleological ordering of the single cosmos. In that context, a teleological explanation looked at least as plausible as any other. Nor ought we to be critical of the Greeks for weak mechanical explanations, since people were struggling to make these work as recently as the seventeenth and eighteenth centuries. A Christian God was often the explanation of the order of the cosmos and the origins of life. It is only relatively recently that we have had a plausible explanation for the origins of life in solely physical and mechanical terms, and there is still a problem within modern cosmology, which splits scientists into 'teleology' and 'multiple worlds' camps, echoing the divisions of the ancients.

6 Medicine and the Life Sciences

Healing is a matter of time, but it is sometimes also a matter of opportunity.

Hippocrates, *Precepts*, chapter 1
(translated by W.H.S. Jones, 1923)

Virtually all societies have had some form of healing practice, just as virtually all have had some form of technology. The Babylonians and Egyptians had reasonable practices, and even possessed some rudimentary knowledge of the human body. It was the Hippocratics, though, who were the first to insist that all diseases had a natural cause, against the previous view that some, or indeed all, were of a supernatural nature. The first task of the healers in ancient Babylon, before any healing could take place, was to decide upon what sin had been committed, in order that the proper purifications and recompense could be made. This was in marked contrast to the early Greek scientists, who generally considered the cosmos to be an entirely natural place, and the Hippocratics, who launched a generalised attack on magical healing practices.

In ancient Greece, there was no established medical profession, and there was a great deal of competition between the various healing practitioners. There were folk-healers, herbalists, magicians and purifiers, to name but a few. One aim of the Hippocratics was to establish themselves as the professional doctors, the people that any well-informed person would turn to in times of illness. So the attack on the supernatural was crucial in the founding of medical science and a medical profession. The Hippocratics claimed that they were the real doctors – the magicians were mere charlatans. They were very keen to insist that there was a proper science of medicine and that they were its practitioners, unlike their opponents or any '*idiotes*' (the Greek for 'laymen', from which we derive the word 'idiot').

The Hippocratics recorded marvellously detailed case studies, though they were not the first to do so. As the Edwin Smith papyrus (which recorded Egyptian medical practices) shows, the Egyptians also recorded cases quite carefully. However, the Hippocratic case studies were remarkable for their detail, their candour, and the fact that they included negative outcomes as well as positive ones. They attempted to study many sufferers from disease, to see how the disease progressed and how various treatments worked. They were very keen on studying fevers, recording the days on which a crisis would occur that would then determine the course of the disease. They were careful to note such indicators as the patient's posture, skin colour and temperature, the reactions of the eyes and the nature of the sputum, vomit, stools and urine.

Admirable though these case studies were, they must be linked to the role of prognosis in the standing of a healer in the ancient world. The arts of diagnosis and prognosis were very important for the early doctor. As one of the Hippocratic writers tells us:

> *If, when he visits his patients, he is able to inform them not only about the previous and present symptoms, but is also able to tell them about what will happen, as well as give further details they have left out, he will increase his standing as a doctor and people will not have worries about placing themselves under his care.*

They were also cautious about terminally ill patients:

> *By recognising that the patient was going to die, and announcing so beforehand, he would be able to absolve himself of blame.*

Prognosis, in particular, was the way for a doctor to earn his reputation against the competition. A cure, or any effective treatment, was another matter. The resources open to the ancient Greeks were very limited, especially in combating disease.

The Hippocratics believed that the human body was composed of the hot, cold, wet and dry, or alternatively that there were bodily humours – bile and phlegm, or black bile, yellow bile, phlegm and blood. While there was disagreement about the exact constitution of the body, one thing that they were generally agreed upon was that health was a balance of the constituents or humours, and disease an imbalance. They had no idea of bacteria or viruses, and no conception of a disease existing outside the body and invading it. Naïve though this theory of disease was, it was an entirely natural theory which served as a coherent basis for diagnosis, prognosis and treatment. We can see the optimism of the Hippocratics in the passage quoted earlier:

> *Each disease has a nature and power of its own, and none is unintelligible or untreatable . . . whoever knows how to bring about moistness, dryness, hotness or coldness in men can cure* [epilepsy] *as well, if he can diagnose how to bring these together properly, and has no need of purifications and magic.*

The ancients concentrated on keeping in good health by means of exercise and diet (regimen). They also used purges, emetics, baths and blood-letting. This could be linked to their theory of disease. If there was thought to be an excess of one constituent (for example, hot), the treatment would be to reduce this excess until a balance was restored. So the Hippocratic treatise *The Nature of Man* tells us that:

> *If you administer a medicine to a man which removes phlegm, he will vomit phlegm: if you administer one that removes bile, he will vomit bile.*

The Hippocratics also made a careful study of environmental factors. In *Airs, Waters and Places* they relate the occurrence of certain diseases to the prevailing weather conditions and the nature of the fresh water supply.

Finally, the Hippocratics are famous for producing an ethical code for doctors, and for the Hippocratic oath. Their code was very philanthropic, and there was a definite disapproval of avaricious doctors. Here are two examples:

> *In whatever houses I enter, I will enter to help the sick, and I will refrain from all intentional injustice and harm, especially from abusing the bodies of men and women, be they free or slaves. Whatever I see or hear in the course of my practice, or in my life outside my practice, I will never seek to divulge, but I will be silent, and leave these things unspoken.*

> *Do not be too unkind, but consider the wealth of your patients. At times, you will treat people for free, recalling to mind a previous benefaction for your*

present reputation. If there is a chance to treat a foreigner or a poor man, do so fully. Where there is love of man, there is also love of the art of medicine.

The Hippocratic code of practice was important in ancient Greece, where healing procedures were completely unregulated. It also helped to separate doctors from non-doctors, a critical issue for the Hippocratics. Aspects of the Hippocratic code are still influential today, especially in relation to euthanasia. All of these important advances of the Hippocratics can be related to their struggle to establish themselves as doctors in the face of competition from other healers. In each case, the Hippocratics would have been able to say, we have the detailed objective case studies, the theory of disease, the code of practice, and so on. We are the real, professional doctors. Whether in the context of ancient Greece, with its limited resources for treating illnesses, they were any more effective than other forms of healer, is another question. What is certain is that the Hippocratics founded both the idea of the medical profession and the scientific study of medicine.

Ancient medicine had its problems. There were very few effective cures for disease, no anaesthetics, and little in the way of analgesics or antiseptics. In general, the ancients were much better at dealing with trauma than disease. The treatment of trauma is a good deal more straightforward and evident, especially if one is just beginning the science of medicine. There was considerable practical experience of dealing with the treatment of battle casualties. The Hippocratics gave very detailed accounts of how to treat various types of wound to different parts of the body, and how to reduce fractures and dislocations. It was in the treatment of trauma that the ancient doctor really could make a difference. The height of this practice was probably reached by the battlefield medics of the Roman army, who in the first few centuries AD were highly organised and efficient, very skilled at treating cuts and amputations – so much so, that they proved better than any other army until the nineteenth century.

The ancients were better at treating trauma to the extremities than to the torso. One major problem was that they had very little idea of the internal functioning of the body, and there was a social and religious taboo against the dissection of humans. Trying to work out the functioning of the organs of the human body from scratch is by no means easy, and the ancients often went seriously astray in this matter. Aristotle, for instance, believed that the heart was the central organ of sensation, while the brain was just there in order to cool the blood! Complicating things further was the fact that there was only a brief period during which both dissection and vivisection of humans was permitted. The later Roman physician Celsus (*fl.* 40 AD) tells us that:

It is in the internal parts that pains and diseases come about, and they believe that no one who is ignorant of these parts can administer remedies for them. Therefore it is necessary to open the bodies of the dead and to examine the viscera and intestines. This was done in by far the best way by Herophilus and Erasistratus who opened men while they lived, men received from the king out of prison, and while these subjects still breathed, they observed parts which nature had previously concealed, their position, colour, figure, magnitude, order, hardness, softness, smoothness, what they touch, the advances and retreats of each, and whether any part is inserted in another or is received by another.

However, this period did not last long. There were also debates about the efficacy of vivisection and dissection. Some thought vivisection cruel and unlikely to provide any benefit. Some thought dissection useless, since they adhered to the motto: 'To heal the living one must study the living.' There were also debates about whether anything could be learned from the vivisection and dissection of animals, though not about the morality of such practices. Vivisection and the use of animals in research was taken for granted in antiquity, and was not a matter of debate. The ancient Greeks and the Romans simply did not share – and indeed had no conception of – our twenty-first-century attitudes towards the rights of animals.

Two key medical thinkers after Aristotle were Herophilus of Chalcedon (*fl.* 270 BC) who was very important in developing anatomy (the study of the structure of the body), and Erasistratus of Chios (*fl.* 260 BC) who was crucial to the development of physiology (the study of the function of the body). Herophilus made important discoveries about the brain and nervous system, and made the first clear distinction between arteries and veins. Erasistratus examined the organs and found each supplied with arteries, veins and nerves, which sub-divided beyond the limit of human perception.

Galen

The most famous doctor in antiquity, and the most influential anatomist, was Galen of Pergamum (*c.* 129–200 AD). Typical of the Hellenistic period, he was a great synthesiser and systematiser. He combined previous medical and anatomical knowledge with his own research to produce the most comprehensive system of medicine and anatomy in the ancient world. With additions and amendments from Arabic culture, Galen's work lasted up until about 1550. At various times, Galen was physician to the Roman army, the gladiators and the Roman emperors. Whatever his experience of the human body in the amphitheatre, Galen faced one major problem. Access to human bodies for dissection at his time was virtually impossible due to social taboos concerning the dead body. Galen said that:

> *It is possible to see human bones. I have done so many times, when a grave has been broken open. Once a river, engulfing a recent hastily constructed grave, easily dissolved it, and with the power of its motion washed the dead body away completely. The flesh had rotted, but the bones were still held in the exact relations to each other.*

Galen thought that if one could get the opportunity to dissect a human body, which was not everyone's luck, one should certainly take it. Occasionally the bodies of enemy war-dead were dissected. One should work on apes otherwise, in order to familiarise oneself, as far as possible, with the position of the internal parts of the body.

It is tragic that such a brilliant anatomist as Galen had so little opportunity to examine the internal workings of the human body. He was an excellent observer of both the human and the animal body, and a gifted experimentalist, and was keen to emphasise the importance of first-hand experience in these matters. Galen was a highly systematic observer, dissector and vivisector of animals. He refuted the idea of Erasistratus that the arteries contain air by tying off arteries above and below the point where he cut them, showing them to be full of blood instead. The methodical nature of his studies can be seen in the following experiment, which was also rather gruesome, as it involved the vivisection of a pig. Galen was trying to find out which parts of the body were controlled by the nerves leading off the spine. To do this, he took a pig and severed the spinal column at each vertebra going upwards, observing which functions the pig lost with each incision.

A good deal of Galen's importance lay in producing a systematic and coherent account of medicine, anatomy and physiology. The Hippocratics had the idea of health as a balance and disease as an imbalance, but now Galen clarified this idea and formalised the humoural system of the body.

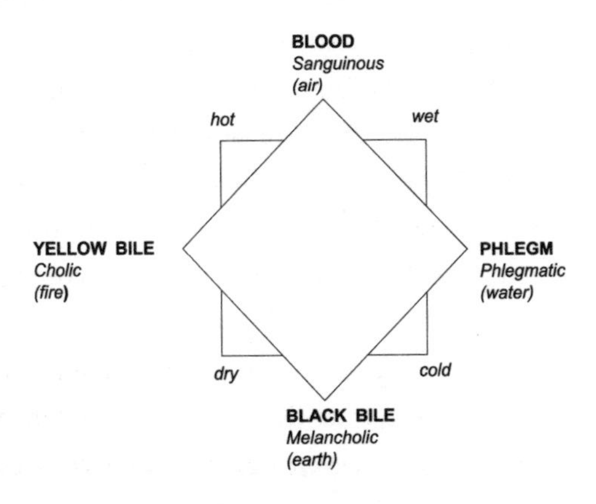

Figure 29: Galen's Humoural System. Associated Aristotelian elements are shown in brackets. There were believed to be four key humours to the human body: blood, black bile, yellow bile and phlegm. Health was thought of as a proper balance of these four humours. Disease occurred when there was an imbalance.

The treatment of disease was an attempt to rebalance these humours. This might be done by diet, exercise, administering purgatives, diuretics or emetics, or by blood-letting. It is here that we see the basis for this ancient and medieval practice. Some patients were considered to be suffering from an excess of blood, thus upsetting the balance of the four humours. They might have some of the symptoms of high blood pressure – red face, protruding veins, etc. The cure was obvious: relieve the body of some blood, and restore the balance of the humours. Similar ideas related to other diseases. With a cold, one has an excess of phlegm; with an infection, an excess of yellow bile (i.e., pus); with coughing up blood, an excess of black bile. The humours had to be brought back into balance.

Galen's work on anatomy was both brilliant and fundamentally flawed. His acute observation and attention to detail allowed him to formulate the most extensive and systematic account of the human body in antiquity. The basic flaw in his work stemmed from the fact that he was unable to dissect a sufficient number of human bodies. As a substitute, he dissected a great number of the higher mammals, in particular Barbary apes, which he considered in many ways to be the animal most like man. Galen was aware of the dangers of supposing the higher mammals to be more like man than they actually are, but unfortunately he still fell into that trap. This was not discovered until the sixteenth century, when dissection in order to generate knowledge was taken up again. The study of anatomy stagnated after the fall of the Roman empire, and although some dissections were done, they were demonstrations to show students that Galen was right, rather than investigations in their own right. This was changed by Andreas Vesalius (1514–64), who began to dissect bodies for himself (rather than having a helper do it for him) and examine them with a more critical eye. This was part of the new Renaissance optimism that tried to outdo the achievements of antiquity. Vesalius recognised, for instance, that Galen's description of the muscles of the human hand was good but slightly faulty. It was, in fact, a brilliant description of the hand of a Barbary ape. Vesalius dissected humans and apes to show that Galen had drawn some of his material from a study of apes. There was a considerable effort in the sixteenth century to purge anatomy of these errors, though Galen's general scheme of anatomy and physiology remained intact.

Ancient Thought on Blood

The idea of the circulation of the blood was not formulated until the work of William Harvey (1578–1657) in the seventeenth century. Galen believed that there were two quite separate systems for the blood, and that the blood did not circulate around the body. Rather, it was slowly generated by the liver and then transported to various parts of the body, where it was consumed. There was a reasonable basis for this belief. Arterial and venous blood are different colours, scarlet and purple. Furthermore, the arteries are thicker than the veins, and carry a pulse. It is not evident to the naked eye that there is a link between the arteries and the veins (this can be seen only with the microscope). If scarlet arterial blood and purple venous blood were in the same system, then there was a need for a process which converted one to the other and vice versa. There was no idea of the role of oxygen and how it affects the colour of the blood, even though Galen realised that something important happened in the lungs with air and blood. In view of these facts, it was quite reasonable to believe that the two types of blood occupied two different systems of blood vessels.

For Galen, the venous system carried the 'nutritive' (purple, deoxygenated) blood. The stomach produced a nutritive, milky substance called chyle which was passed to the liver, where nutritive blood was gradually generated. The venous system radiated out from the liver, the seat of nutrition. The veins carried the nutrition to the rest of the body, some (but not all) of this blood passing through the right side of the heart. The arterial system carried 'vivified' (scarlet, oxygenated) blood. This system originated in the lungs, which vivified the blood, and then carried this life-giving spirit to the rest of the body via the left side of the heart (see Figure 30).

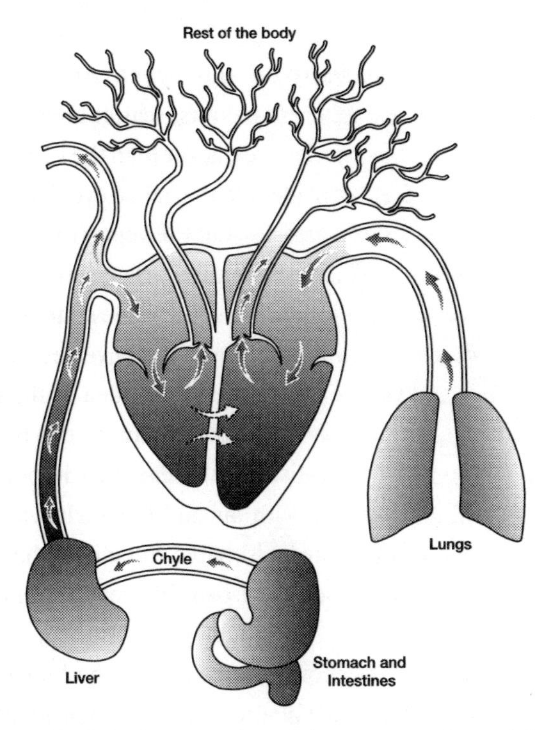

Figure 30: Galen's conception of the venous and arterial systems.

In both systems, blood did not return to its point of origin, but was gradually consumed by the body. As the arterial blood was consumed, it was necessary that some venous blood be transported into the arterial system and turned into vivified blood. There was believed to be a slight seepage from the right side of the heart to the left (the venous to the arterial) via the septum, in order that the arterial blood could be replenished. In fact, there is no flow across the septum in healthy humans after birth, although it does occur in the foetus.

Galen's account of the heart supported his views on the motion of the blood. The motion of the heart is actually extremely difficult to analyse, since it is very swift and it is by no means clear what is actually happening inside the heart. Galen believed that the active phase of the heart, when the muscles operated, was when the heart expanded in volume, so that the heart attracted blood into itself. The heart became smaller as the muscles relaxed, so that blood was not expelled from the heart with any great force. Actually, this is the wrong way around – the active phase of the heartbeat is the contraction. This was associated with another error on Galen's part. For him, the active phase of the heart did not coincide with the pulse, and so he did not believe that the heart caused the pulse in the arteries. He believed that they pulsed of their own accord, and attracted blood into themselves. He thought that they were rather like the intestines. During vivisections, Galen had observed peristalsis, the process by which the gut moves food along itself by contractions. He believed that the gut attracted food into itself by this process. The pulse, he believed, was the way in which the arteries attracted blood into themselves. Attraction was a very important general principle of Galen's physiology. Things were not forced around the body – organs attracted what they needed to themselves.

Galen's whole account of the cardiovascular system supported the idea of a slow flow of the blood. The blood was produced slowly, distributed slowly and consumed slowly. He could quite easily account for rapid blood loss when someone was cut, and even arterial spurting (when arteries are cut, blood is thrown out in vigorous spurts). Blood can be under pressure, and so flow rapidly from a wound, without moving rapidly around the body, while the arteries would spurt because of the pulse they generate themselves.

Harvey's discovery of the circulation of the blood was important not only as the correct account of the distribution of blood around the body, but also as a significant blow against Galen's physiology. Harvey discovered that the blood flowed rapidly around a full circuit of the body, that the active phase of the heartbeat was a contraction such that blood was expelled from it with considerable force, and that the pulse was due to the heart and not the arteries themselves. This dealt a severe blow to the principles of attraction that were so important to Galen.

Galen had a coherent and comprehensive account of human anatomy, physiology and medicine. His views could be used as a basis for diagnosis, prognosis and treatment. He also gave a plausible account of the assimilation of food, the production of blood, the distribution of nourishment to the body, the heartbeat and pulse, and the production and distribution of heat around the body.

Aristotle and Biology: Biology's Beginnings

While it was Galen's work on the human body that was the bequest of antiquity, it was Aristotle and his followers who set the agenda in biology. There was no real study of either the animal or plant kingdoms prior to Aristotle. A standard criticism of Aristotle and other ancient scientists is that they engaged in too much theorising and not enough experiment and observation. This is simply untrue of Aristotle's biological work. He was quite remarkable in founding an observational study of plants and animals. He spent time travelling throughout Greece, becoming acquainted with 500 species of animals, and making many careful and detailed dissections. His attitude to this study can be judged from the following:

> *We should not feel a childish disgust at the investigation of the meaner animals. For there is something of the marvellous in all living things.*

Aristotle had a considerable struggle to get biology accepted as a proper field of study for the scientist or philosopher. The Greeks held a strongly hierarchical conception of nature and the cosmos, such that the study of its finer parts, such as the heavens or human beings, was accepted, while the investigation of its lesser parts was seen by some to be demeaning. Aristotle's views were very important in getting the study of animals and plants onto a serious footing. Central to this was Aristotle's view that nature does nothing in vain, that all animals are well organised and an expression of the good.

Embryology and Species

Aristotle made a very detailed study of the development of animal embryos. He supported the idea of 'epigenesis' against those who believed in pre-formation. Epigenesis was the view that the embryo develops its different parts from an amorphous beginning, rather than having all of its parts ready formed. He argued against 'pangenesis', and for the fixity of species. Pangenesis was the idea that characteristics acquired during life are passed on to the offspring, e.g., if giraffes stretch their necks to feed, then baby giraffes are born with longer necks. Aristotle also argued against the ancient idea that mutilated humans produced deformed offspring. His view was that the male provides the form, the female provides the matter, and the embryo has then a potential to become an adult human, which it proceeds to actualise. Inherent in this is the idea of the fixity of species – that there is no evolution. Humans (and all other species) are what they always were (and will be), and the boundaries between species are permanent.

Taxonomy

A serious problem for biology in the ancient world was how to arrange species of animals into groups. Aristotle produced the first system to give a reasonable classification of animals. Rather than basing his system on organs of movement, as previous attempts had done, Aristotle used modes of reproduction. The main groups that he devised were: the viviparous, which bear live offspring; the oviparous, with perfect or imperfect eggs; and the insects, with larvae. Animals were then divided up into genus and species. The system was a hierarchy of perfection, man naturally being considered the most perfect species. The modern system of classification dates from Linnaeus in the eighteenth century.

Theophrastus (371–286 BC) followed Aristotle as head of the Lyceum, and continued his biological work, classifying many species of plants. Many Greeks believed in the spontaneous generation of both plants and small animals (as did many people up to the eighteenth century). Theophrastus was more reserved about this, arguing that the wind carried many small seeds which could account for the generation of plants. He was also the first to make a comprehensive study and classification of rocks. Against Aristotle, he argued that there were limits to the expression of the good in nature. It seemed to him that some parts served no function (such as human male breasts), while some parts could be better arranged and some were even harmful. He did not reject teleological arguments outright by any means, but he did realise that there were limits to this sort of argument, and that many things in nature did occur by chance.

7 Later Greek Science: After Aristotle

. . . Nil posse creari
De nilo
(Nothing can be created out of nothing.)
Lucretius, *De Rerum Natura*, book 1, l. 155

Later Greek science and philosophy is usually taken to start with the death of Aristotle in 322 BC (and Alexander the Great in 323 BC), the beginning of the era known as the Hellenistic period. The major periods for ancient science were:

Babylonian – 1000 BC onwards
Pre-Socratic – 600–400 BC
Athenian – 400–300 BC
Hellenistic – 300 BC–200 AD
Roman – 200–600 AD

The Hellenistic period was marked by its syntheses and accumulation of knowledge in certain spheres of learning, most notably by Ptolemy in astronomy, Euclid in geometry and Galen in medicine. Ptolemy gathered together existing knowledge in astronomy and, together with his own contributions to the subject, he synthesised a comprehensive new system that lasted until about 1600. We see a similar synthesis and longevity with Galen in anatomy, physiology and medicine, and with Euclid in geometry, though Euclid's work lasted longer. The Hellenistic period saw several rival groups of philosophers vying with one another. Plato's Academy and Aristotle's Lyceum were still going strong, and both philosophers had their adherents. The other groups with important views about the natural world were the Epicureans and the Stoics.

Epicurus and Epicureanism: on the Nature of Things

Epicurus of Athens (*c*. 342–271 BC) was an atomist in the tradition of Leucippus and Democritus. Aristotle, as we have seen, did not believe that matter came in small, discrete particles. Epicurus believed that only atoms and the void existed. Atoms came together to form bodies, and all that we perceive (hotness, colour, etc.) could ultimately be explained in terms of the motion and the arrangement of these atoms. Aristotle had criticised Leucippus and Democritus for not distinguishing between physical indivisibility and mathematical indivisibility. However small a physical atom was, one could imagine something smaller simply by imagining something half the size. Epicurus stated very clearly that the atoms were physically indivisible but could be divided in thought into indefinitely small mathematical parts. He also placed some limits on the sizes of atoms, whereas Leucippus and Democritus had believed that they came in all shapes and sizes – though this had the unfortunate consequence of implying that there were atoms so large as to be visible. Leucippus and Democritus also believed that the world had come into being from a vortex, and that this vortex formed spontaneously. Epicurus, on the other hand, believed that all of the atoms were moving in one direction ('down'), and on parallel paths. All atoms moved through the void at the same speed. Occasionally, an atom would swerve out of its path and so interact with other atoms. It was in this manner, Epicurus thought, that worlds began to be formed.

The ultimate goal of the Epicureans was happiness. They rejected the view of Plato and Aristotle that one should lead a life devoted to the good, and instead aimed at pleasure. This affected the depth to which they were willing to analyse problems. Once they had an answer that they were happy with, they went no further. In some fields they went reasonably deep, while in others they barely skimmed the surface, professing themselves happy with the state of things. Epicureanism, for all its faults (and the idea of the inexplicable atomic 'swerve' was severely criticised, even in antiquity) was a significant trend in Hellenistic times. It was influential, too, in the Roman world – Lucretius (94–55 BC) wrote an epic poem giving an atomistic view of the world. The great problem for ancient atomic theories which relied on mechanism and chance was that they lacked the resources to produce plausible explanations of the phenomena, whether one is talking of cosmogony and cosmology or more mundane areas. They lacked any convincing account of the way in which atoms were brought together, or how they stayed together. Effectively, they were brought together by chance and stayed together by chance, hence the need of the atomists to postulate multiple worlds. The teleological accounts of Plato and Aristotle must have seemed at least as plausible at the time.

Stoics: the Active and the Passive

Another important school were the Stoics, whose founders were Cleanthes of Assus (331–232 BC), Chrysippus of Soli (*c*. 280–207 BC) and Zeno of Citium (335–263 BC) – not to be confused with Zeno of Elea. The Stoics, like the Epicureans, held that the main goal was to be happy, but they worked out an account of the cosmos that was far more detailed than anything offered by the Epicureans. Unlike them, the Stoics denied that there were atoms and a void. The cosmos was a plenum – that is, every space was filled – although the cosmos itself was situated in a void which surrounded it. The Stoics held that there were continua which could be indefinitely divided, and so there were no atoms. Like most of the ancient Greeks, they believed that there were four elements: earth, water, air and fire. They also held that there were two fundamental principles: the active and the passive. The passive was associated with inert, quality-less matter, while the active principle, or '*pneuma*' (breath), was associated with god. Fire was thought to be the most active element, also associated with god. It is important to stress that both of these principles were corporeal, and that they came together in what the Stoics called a '*crasis*', or total mixture. The Stoics were therefore pantheists, since the whole of their cosmos was in a sense permeated with god, and was god.

The Stoic account of the origin of the cosmos held that initially everything was a universal conflagration or '*ekpyrosis*'. This fire gradually changed into air, and then to the other elements. There was also supposed to be a converse process whereby the elements turned back into fire. Thus, the Stoics held a cyclical conception of the cosmos – it had no origin in the strict sense for them. In the part of each cycle in which the cosmos was coming into being from fire, it was guided by god, who retained the '*spermatikoi logoi*', the seminal principles, through the conflagration. The Stoics disagreed with the atomists on whether the cosmos came together by chance, and their cosmos was permeated not merely by intelligence but also by providence. God had a plan for the cosmos, and it was a good plan. In these senses, the Stoics were much closer to Plato and Aristotle than to the atomists. They also held a strongly determinist view of the world. Everything that happened was pre-determined, and what is more, it would happen again in the next cycle of the cosmos. So according to the Stoics, you have already read this book in a previous cycle of the world, and you will be reading this book again in the next cycle.

A very important idea, both in Stoic cosmology and in Western thought prior to the seventeenth century, was that of the macrocosm/microcosm relationship. This is the idea that the cosmos on a large scale – the universe – functions in the same way as, or has structural similarities to, the cosmos at small scale – human beings. So for the Stoics, the cosmos was a living creature pervaded with the active principle *pneuma*, and had intelligence. Each of those things could be said of the microcosm – humans – as well. This idea of an organic unity to the cosmos, a relation between the macrocosm and microcosm, was immensely important for Western thought until the scientific revolution.

Archimedes

Archimedes of Syracuse (287–212 BC), the son of an astronomer, provides the title for this book. It is often said that he leapt from his bath shouting '*Eureka*!' ('I have found it', actually '*heureka*' in ancient Greek), and then ran home naked, having solved a problem that was perplexing him. The problem was whether the crown of King Hieron II, supposed to be pure gold, was indeed so. It weighed the same as the gold delivered to the goldsmith, but had he adulterated the gold with silver and made a fraudulent profit with the excess? Archimedes, who had made a careful study of hydrostatics, came up with the following solution. Take an amount of pure gold equivalent in weight to the crown, and measure how much water this displaces. Do the same with silver. If the crown displaces more water than the equivalent weight of pure gold, then it has been adulterated in some way. The tale has it that Archimedes realised this while lowering himself into his bath. While this is a splendid tale, we have no proper evidence for it. There is a series of such apocryphal tales in the history of science. There is no evidence for Archimedes leaping from his bath, no evidence that Galileo dropped cannon balls from the leaning tower of Pisa (he knew of much better experiments than this already), no evidence that Newton had a realisation about gravity while sitting under an apple tree, or that Watt invented the steam engine while watching a kettle boil (the steam engine had long been invented, and Watt's improvements to it had nothing to do with steam expanding).

Archimedes was a brilliant mathematician and engineer. His work in geometry, his true love, developed the work of Euclid. He tended to look down on his engineering work. Cicero (106–43 BC), the Latin poet and philosopher, tells us that Archimedes refused to write any practical treatises, confining himself to theory and mathematics. Cicero also tells us that he constructed a mechanical model of the heavens which

> [W]*ith a single motion reproduced all the unequal and different movements of the heavenly bodies.*

Archimedes is probably most important for his work in mechanics and in hydrostatics. He developed the theory of the lever, and formulated the principle that

> *Two weights balance at distances reciprocally proportional to their magnitudes.*

He also recognised that, in principle, one could move very great weights with a relatively small force if one had a large enough lever, or a similar means of multiplying forces, and so he said:

> *Give me a place to stand and I will move the world.*

It is said that Archimedes gradually pulled a large ship ashore using a system of pulleys to multiply forces, to the astonishment of those present. His study of the properties of fluids, and whether objects float, was also of great importance in antiquity. He formulated the important principle that:

A body immersed in a liquid loses weight equal to the weight of the liquid displaced.

There are sources which tell us that Archimedes invented the water screw, a device for raising water, and the compound pulley, a means of arranging pulleys in order to multiply force. It is unlikely that he actually originated either of these devices, but he may well have improved them and given an account of the principles involved, and how to use them in an optimum manner.

Archimedes is said to have helped in the defence of Syracuse by inventing military engines. The most famous of these is his 'claw', by which it is claimed that the defenders could upend ships approaching the sea walls and sink them. Unfortunately, we do not know the nature of the claw, and can only speculate on how it worked. It is likely to have been a device whereby means for multiplying forces, such as levers and pulleys, were used to lift one end of a ship when grasped by the claw. The other end of the ship would then dip below the water, and it would sink rapidly. Archimedes died when Syracuse was finally sacked by Roman troops in 212 BC.

Eratosthenes

Eratosthenes of Cyrene (*c.* 276–195 BC) is famed for his remarkably accurate estimation of the size of the earth. He knew that at noon on the day of the summer solstice, a rod placed in the ground at Syene (near Aswan in Egypt) cast no shadow, and a well was fully illuminated at its bottom, so the sun was directly overhead. Yet a rod at Alexandria cast a shadow of 1/50 of a circle, just over 7°. A simple piece of geometry then told him that the distance between Alexandria and Syene was 7/360 of the earth's diameter. The distance was estimated at 5,000 *stades*. A '*stade*' was originally one lap of a stadium, and we believe that to have been 157.5 metres, so we get an estimate for the diameter of the earth which is 39,690 kilometres, remarkably close to the modern figure of around 40,000 km. Admittedly, we are not quite sure

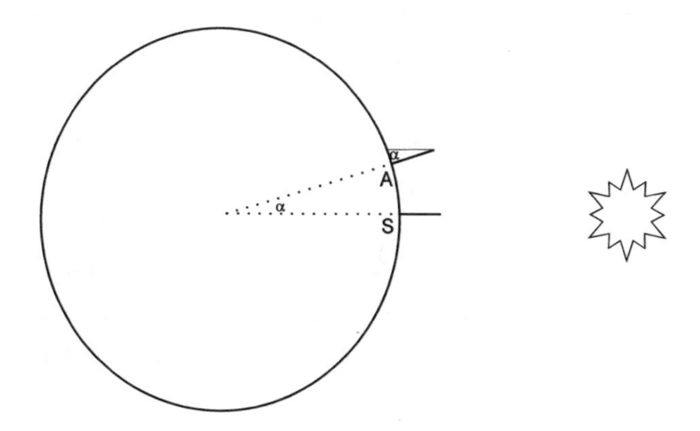

Figure 31: Eratosthenes and the size of the earth. The angle cast by a rod at Alexandria when a rod casts no angle at Syene is equal to the angle between them, if they have the same longitude. If the distance between them is known, then one can calculate the circumference of the earth. Not to scale!

how long Eratosthenes' *stade* was, but even if he was using any of the other possible lengths, his calculation would still be reasonably close, within around 15 per cent. The Greeks not only knew the earth to be spherical, they also had a very good measure of its size. Eratosthenes produced very good work in geography, which was later used by Julius Caesar.

Hero and his Engine

One of the great inventors and technologists of later antiquity was Hero (or Heron) of Alexandria (*fl.* 60 AD). He did something quite remarkable. He invented a steam engine with a rotary motion 1,500 years before Watt and the industrial revolution. In all fairness, Hero's engine was a pretty crude affair, and to construct a steam engine that would have produced any meaningful amount of power was far beyond the technological capacity of the ancients. In Hero's engine, steam was generated in a large cauldron and then passed into a small rotating sphere with exhaust pipes pointing in opposite directions, thus generating a rotation of the sphere. The rotation was not very powerful, but there would have been movement, nonetheless. Proper steam engines actually work by another principle. If a cylinder full of steam is rapidly cooled, the steam condenses, and water and a vacuum are produced. A piston is then moved by the pressure difference between the inside of the cylinder and the outside, producing power. So Hero's engine was a long way from a proper steam engine (and this, incidentally, is why Watt watching expanding steam from a kettle has nothing to do with his innovations).

Hero also invented many other toys, including temple doors which opened by steam power. The Greeks were inventive, and their technology did progress. In particular, devices relating to warfare improved, as did methods of producing food. Overall, though, they failed to establish a fruitful relationship between technology and the sciences at a general level.

Figure 32: Hero's steam engine.

The Origins of Alchemy and Astrology

There were two other disciplines that had a considerable influence over Western thought prior to the scientific revolution, and whose genesis can be traced back to ancient Greece. These were astrology and alchemy. Why mention these in a book on science? Firstly, in the ancient world there were significant thinkers who considered both astrology and alchemy to be sciences. Secondly, ancient astrology and alchemy were significantly different from their modern counterparts, as were ancient cosmology and matter theory. It is possible that both ancient astrology and alchemy had a different relation to the science of their day.

Ancient astrology and alchemy were both considerably broader in their conception than they are generally given credit for, and there was a wide spectrum of views on how they might work. Alchemy was not merely the search for the transmutation of base metals into gold. Rather, it dealt with the transformation of less valuable things into more valuable things in general. Astrology dealt with the influence of the heavens on the earth in general, and not just on human beings. The greatest of the ancient astronomers, Ptolemy, who was also hugely influential in the history of astrology, set down the evidence in favour of astrology. He began by mentioning the effect of the motions of the sun on the seasons and on the seasonal behaviour of animals and plants, and the effect of the moon on the tides and on the nocturnal behaviour of animals and plants. These effects were undeniable, and were taken to be part of astrology. Only later did Ptolemy move on to the effects that the heavenly bodies may have on humans. Ptolemy's book the *Tetrabiblos* was enormously influential in astrology until the seventeenth century.

The origins of astrology followed the lines that we have seen for astronomy and medicine. The Babylonians and the Egyptians both studied the heavens and produced astrological portents. It was the Greeks who provided a theoretical basis for astrology. Aristotle himself was not interested in astrology, but his cosmology was used by many who were. The key point was that Aristotle believed that the four elements would have separated out into concentric rings of earth, water, air and fire, had it not been for the action of the sun in stirring up these elements. There was a sense, then, in Aristotle's cosmology, that the sun was responsible for everything that happened on earth. Clearly, the sun had a heating and drying effect. In addition to this, some supposed that the moon had a moistening and cooling effect. Since hot, cold, wet and dry were the pairs of primary contraries which underpinned earth, water, air and fire, the sun and moon could affect anything on the earth, including perhaps human beings. The means for the transmission of astrological influence therefore depended only on Aristotle's cosmology and his matter theory. It is important to stress that there was nothing mystical or supernatural about Aristotle's cosmology or his theory of matter. Ancient astrology done in this manner could meet one of the key modern objections to astrology. This is: how do the heavenly bodies affect the earth in a significant manner, without falling back onto anything implausible, supernatural or mystical? Indeed, ancient astrology could meet many of the standard objections to astrology, at least in principle. The fact that astrology was construed so broadly meant that some of the predictions (the more meteorological sort, or those pertaining to the seasonal or nocturnal behaviour of animals) could be quite precise, instead of vague and untestable. So there seemed to be empirical evidence in favour of astrology, when conceived this broadly. Astrology, as formulated by Ptolemy, was not based on outdated astronomy and cosmology. It was based on Ptolemy's latest discoveries in these fields. Astrology had not yet proved itself incapable of making any progress, and to many it seemed as plausible a science as many other ancient sciences which were still being developed.

One can see why, in an ancient context, some people would have believed astrology to be a science. There were also many who objected, and there were fierce debates about whether astrology worked at all, and whether it could tell us anything about humans. Not all astrology was done in an Aristotelian manner. Plato's philosophy could also be used to underpin astrology, and there were many other possible bases. There was a broad spectrum of views on how astrology might work. At one end we have the Aristotelian interpretation, requiring nothing mystical or supernatural. The Stoics, with their strong belief in determinism, were also interested in astrology. Moving along the spectrum, the macrocosm/microcosm relationship could also be used to support astrology, if changes in the universal, macrocosmic mind (i.e., the heavens) were reflected in microcosmic minds, those of humans. At the far end of this spectrum were mystical or supernatural ideas on how astrology might work.

Alchemy had a slightly different status to astrology in the ancient world. While there was considerable debate about whether astrology was a science, alchemy was more generally accepted. The origins of alchemy are similar to those of astrology. There was a long tradition of metal working in Egypt, and a good deal of practical knowledge. The Stockholm and Leyden papyri show the Egyptians to have been interested in the production of gold, silver, jewellery and dyestuffs. Again, the Greeks provided the theoretical basis for thinking about alchemy. If we ask how people thought alchemy might work, then we need to go back to Aristotle again, this time to his theory of matter. As we have seen, Aristotle believed there to be four elements: earth, water, air and fire; and two pairs of contrary qualities which underpinned these elements: hot and cold, wet and dry. These elements were by no means fixed. Transmutation could happen quite easily, and without recourse to anything magical or unnatural. If water (cold and wet) was heated, it became air (hot and wet). Aristotle believed all things to be made of the four elements. Metals, including gold, were a combination of earth and water. If the right process could be found to alter the proportions of hot, dry, wet and cold, then one substance could be changed into another. Aristotle also believed that metals were generated in the ground. Thus, there was a process by which metals, including gold, were formed from other substances. Again, for Aristotle this was an entirely natural process, similar to the way in which we might say that coal forms in the ground. An alchemist might quite reasonably hope to replicate and perhaps accelerate this process.

Many other processes which increase the value of something were thought of in this way as well, e.g., the production of dyestuffs. So ancient alchemy could be conceived of entirely within the framework of Aristotle's theory of matter, about which there was nothing mystical or supernatural. This theory was hugely influential in the ancient world. Many alchemists did think about alchemy in precisely these terms. We can therefore see why, in the ancient world, some people would have been happy to call alchemy a science. As with astrology, there were many ways in which one could theorise alchemy. These ranged from a strict Aristotelian basis, to other theories of matter, including the active and passive principles of the Stoics, to outright mysticism and supernatural ideas.

The Decline of Greek Science? All Good Things Must Come to an End

Science in the ancient world eventually went into a decline, although not until the later stages of the Western Roman empire. Certainly, science was still going strong in the first and second centuries AD, with the works of Ptolemy and Galen. There were important later thinkers such as Philoponus, who produced criticisms of Aristotle, Simplicius, a commentator on Aristotle's scientific works, and Iamblichus, a follower of Plato who stressed the importance of mathematics in science. However, the general trend in the latter part of antiquity was towards less creativity and activity in science. One reason for this may have been the later fragmentation of the Roman empire, leaving less time and fewer resources to investigate either philosophy or the natural world. Another reason may have been the rise of Christianity. While early Christianity was not uniformly hostile to science, there were certainly powerful tendencies to look to the spiritual rather than the physical, and to do away with pagan – and in particular Greek pagan – thinking. So Tertullian (*c.* 160–225 AD) said:

> *We need not be afraid if the Christian does not know the powers and the number of the elements, the motions and eclipses of the heavens, the nature of the*

animals, plants and stones . . . It is sufficient for the Christian to believe that the cause of everything created, whether in heaven or on earth, visible or invisible, is the goodness of the Creator, of the one true God.

Slightly later, in 390 AD, we can find Augustine (354–430) saying:

What has Athens to do with Jerusalem, the Academy to do with the Church, the heretic to do with the Christian? . . . We have no need for curiosity after Jesus Christ, and no need of investigation after the gospel. Firstly we believe this, that there is nothing else that we need to believe.

Hermias at least displays some sense of humour when he says of the Stoics:

Will you listen to the nonsense their philosophers speak, when they say that fire is God? They mistake the deity for their destination.

The Creation of Science

It was, then, the ancient Greeks who were the originators of science, although not, it must be said, without considerable contributions from other cultures. But it was the Greeks who took the technologies of earlier civilisations, most notably those of the Babylonians and the Egyptians, and turned them into science. Equally, there have been other contributors to the tradition which has led to modern science. The Arabic/Islamic culture did a great deal to preserve, and then extend and transform, Greek scientific thinking during and after the decline of the Roman empire. Much of this was transmitted to the West, helping to rouse it from the dark ages and push it on towards the Renaissance and the scientific revolution. The Romans contributed a great deal of technology, even if their scientific achievements were meagre. Technology and scientific ideas from China, travelling to the West along the trade routes, were also important influences for Western science.

The ultimate origins of science lay with the Greeks, though. They rejected explanation in terms of myths and capricious gods, and considered their cosmos to be an entirely natural and well-ordered place. In distinguishing between the natural and the supernatural, they effectively discovered nature. They began to use theories to describe and understand their cosmos. These theories were couched in natural terms, and importantly could be discussed and improved upon in ways that myths could not. It is the rapid increase in the sophistication of their theories that is perhaps the most remarkable thing about the ancient Greeks. Wherever one looks – in cosmology, theory of matter, medicine – one sees them making huge conceptual leaps and arriving at new, better theories. We can also see the Greeks developing means to resolve debates about theories and being conscious of a distinction between science and technology.

That is what happened among the early Greeks to establish science. We might also ask: why did science begin with the ancient Greeks? Why, in particular, did science begin with the Milesians? To say that there was a Greek 'miracle' which brought about the birth of both science and philosophy would not explain anything, and would ignore one of the key lessons of the Milesians. While the Milesians achieved a good deal, it is important to put their achievement into perspective. They neither created science from scratch, nor produced the finished article. They were considerably indebted to other civilisations, especially the Babylonians and the Egyptians. One of the most important things about the Milesians was not the actual quality of their theories, but that they conceived of the idea of a cosmos as an orderly and natural place which could be explained by theories. This, along with the intellectual and religious toleration of ancient Greece, allowed Greek science and philosophy to develop rapidly. The other pre-Socratics, then Plato, Aristotle and the Hellenistic thinkers carried forward this programme of explaining the cosmos in natural terms. The Greek achievement was the work of many hands over a great period of time. The contribution of the Milesians was good, and was seminal, but it was not miraculous. They themselves would have been the first to argue that.

The conditions which aided the Greeks were their lack of a central religion and hierarchical organisation, and freedom of expression, allied to a society affluent enough for some people to have the leisure time to investigate questions about the nature of the world. In addition, the technological bases in some disciplines (geometry, astronomy, medicine) were already in place, so the time for a transformation to science was ripe. Greek society, with its love of criticism, debate and knowledge for its own sake, proved to be an immensely fertile soil, once the seeds of science had been sown.

Greek science had distinctive strengths and weaknesses. In very broad outline, the great strengths and achievements were almost all intellectual and theoretical, the weaknesses mainly practical. The fact that Greek philosophy and science were so closely intertwined was a double-edged sword. It allowed the Greeks to develop what was so desperately needed – a natural conception of the world about us, and a theoretical framework for the sciences. It also allowed them to break with mythopoeic thought. The other side of this situation is that some, though by no means all, of these philosopher-scientists were interested only in the philosophical and theoretical aspects of science, and neglected the practical aspects.

Undoubtedly, the strongest areas of Greek science were those in which there was no need for observation, experiment or a strong link with technology, or in which, for some special reason, the Greeks inherited a good deal of empirical data or had no objection to gathering such data. So Greek mathematics and geometry were strong, because these disciplines, as conceived by the Greeks, did not require observation and experiment. Even here, though, the more theoretical work of the Greeks grew out of the practical mathematics and geometry drawn from Babylonian and Egyptian sources. Greek astronomy was strong, because it had access to Babylonian and Egyptian records, and observing the heavens was accepted as a dignified pursuit. Greek cosmogony and cosmology was also rich in ideas, if a little short on empirical confirmation of them (though, in fairness, cosmology really became an observational discipline only in the twentieth century). Greek medicine was relatively strong too, again partly because it inherited a good deal of practical knowledge from the Babylonians and Egyptians, and partly because medicine in ancient Greece was a highly competitive business. The Hippocratics needed to develop effective treatments, as far as was possible in the ancient world, and realised that to do so they must make careful observations, experiment with various possible treatments, and make full use of whatever technology of healing was available.

The major weakness in ancient Greek science was a lack of appreciation of the proper role of experiment, observation and technology. Experiment, to some extent, was seen as manual labour, and as such beneath the dignity of a 'gentleman' philosopher. Xenophon (*c.* 430–354 BC), a contemporary of Plato and Aristotle, said that:

> *What are called the mechanical arts are spoken against, and, naturally, are held in utter contempt in the cities. They ruin the bodies of the workmen and overseers, compelling them to be seated and to live in the shade, spending the day at the fire.*

The Greeks were never particularly good at technology (the Romans were much better), nor did they develop a fruitful relationship between technology and science. The prime example is that of Hero's engine. Hero invented a steam engine which could produce rotary motion. A crude, very low-powered engine, but an engine nevertheless. What did the Greeks think of this? Was it a source of power that could possibly be used in many situations? No. It was an interesting toy, a party-piece to impress people with. Because of this sort of attitude, the Greeks never put the necessary developmental work into their technology. They had no real appreciation that a fruitful liaison between science and technology could lead to inventions which would better the lot of society in general, and indeed would improve their science. Perhaps this was due to their slave-based culture (there was no need for labour-saving devices), or the stigma of manual work, or the nature of the aristocratic Greek philosophers. Whatever the answer, this was a weakness of Greek science.

There were also some more specific weaknesses. While the Greeks did many wonderful things in cosmogony and cosmology, there was a fundamental limitation which coloured all of their work. They never developed a conception of gravity. They were forced to explain its effects by means of other theories, and this affected the nature of all of those theories. From the early 'parallel' conception of the cosmos, to the 'centrifocal' conception, to the sophisticated ideas of later antiquity, the struggle to account for gravitational phenomena coloured Greek cosmology. Some of the Greeks believed in a like-to-like principle, while Aristotle's theory of natural place came to be dominant. The Greeks had no proper conception of force as we understand it, nor did they develop the idea of relative motion. A combination of these factors led the Greeks to believe that the earth was immobile. They believed that if it moved there would be fierce winds; they could not see how it would hold together, nor why we would stick to its surface if it was in motion away from its natural place. Geocentrism created a problem for Greek astronomy. All of the motions of the heavenly bodies had to be real ones – some couldn't be merely apparent, and due to the motion of the earth. This meant that the Greeks had to develop complex devices to generate the motions of the planets. While one can understand why the Greeks adopted geocentrism, this remained a weakness in their astronomy and cosmology.

Several of the ancient Greeks emphasised the need for mathematics in their understanding of the cosmos. However, it was by no means evident how mathematics related to the natural world, and the Greeks, except in a few isolated instances, never really employed the idea of mathematically formulated laws of nature. They were also too optimistic about how well-arranged life forms and the cosmos were, and too liberal with their use of teleology. There were reasons for these specific weaknesses, not least of which is that the Greeks were the pioneers of science, and science is not easy. Ideas such as universal gravitation do not come easily, as is evidenced by the fact that it took a further millennium after the end of Greek science for this idea to be formulated. Non-teleological accounts of the cosmos and of life processes require a great deal of sophistication to become plausible, and this has been achieved only in the last few hundred years. The weaknesses need to be placed in context against the great achievements and advances of Greek science, and the remarkable fact that the Greeks managed to get science off the ground at all. That cannot have been easy in a culture which still relied heavily on mythology and a theology of mischievous, interfering gods.

There are, of course, other important differences between modern and ancient science, particularly in terms of organisation and funding. Whereas science nowadays is funded by industry and the state, ancient research was carried out on a purely amateur basis by those with the interest, resources and leisure time to pursue it. Instead of the laboratory, university or research institute, the ancients had at best a crude observatory. The philosophical schools, such as the Academy or the Lyceum, important as they were, gave no support comparable to that which we can find in the modern world. The relationship between science and technology is now much tighter and much better understood, but was considered virtually irrelevant by the ancients. The number of people doing science in the ancient world was proportionally far smaller than nowadays, and it is always wise to remember that the history of Greek science is the history of a small but influential group of thinkers. Finally, we might compare the instruments available to the ancients with those nowadays. They had a few crude instruments for observing the heavens, but little else with which to investigate their world – no microscopes or telescopes, no thermometers, and only relatively primitive means of measuring distance and weight. Modern science depends on precision instruments to investigate accurately. The ancients were virtually devoid of these.

When summing up the work of the ancient Greeks, it is important to remember that prior to the scientific revolution of the seventeenth century, there was little reason to suppose that the mechanical and atomistic world view that came to prominence at that time was correct. There were many other possibilities, and because of the problems that mechanical and atomist accounts had in explaining how the order of the cosmos came about, and how life originated, the teleological accounts of Plato and Aristotle would have seemed at least as plausible to the ancients. In general, atomist and mechanical accounts were weak in antiquity, lacking as they did many modern resources. One result of this was some ancient attitudes to explanation. The ancients tended to have a more organic conception of the cosmos, use more organic metaphors and explain more holistically than we would, since, for them, this seemed a more convincing approach than attempting mechanical and atomistic explanations. It is also important to recognise the length of time involved with Greek science. We are talking of nearly a millennium, from around 600 BC to 200–300 AD. We would be very cautious in considering the science of the period from 1200 to the present day as one entity, and drawing general conclusions from it. We should be duly cautious of doing so with the Greeks as well. This is especially so given the diversity of thinkers and ideas that proliferated in ancient Greece.

Having said that, Greek science stands as one of the great achievements of the ancient world, and indeed one of the great achievements of humankind. To have begun science is remarkable enough in itself, but especially so in the ancient world. The vision and clarity of purpose of the ancient Greeks was exceptional, as was their tenacity in pursuing the view that the world is comprehensible and can be explained in a rational manner. The way in which Greek theories increase so rapidly in sophistication once they get the elements of science in place is astounding, and is comparable to any of the great periods of human intellectual endeavour, such as the Renaissance, the scientific revolution and the Enlightenment. They bequeathed a wealth of fascinating ideas and arguments about the nature of the physical world, many of which have been important in the development of science, and some of which are still relevant today. Above all, though, they gave us the basic structures and vision of science. That is something that stays with us as we find out more and more about the cosmos that we live in.

Appendices

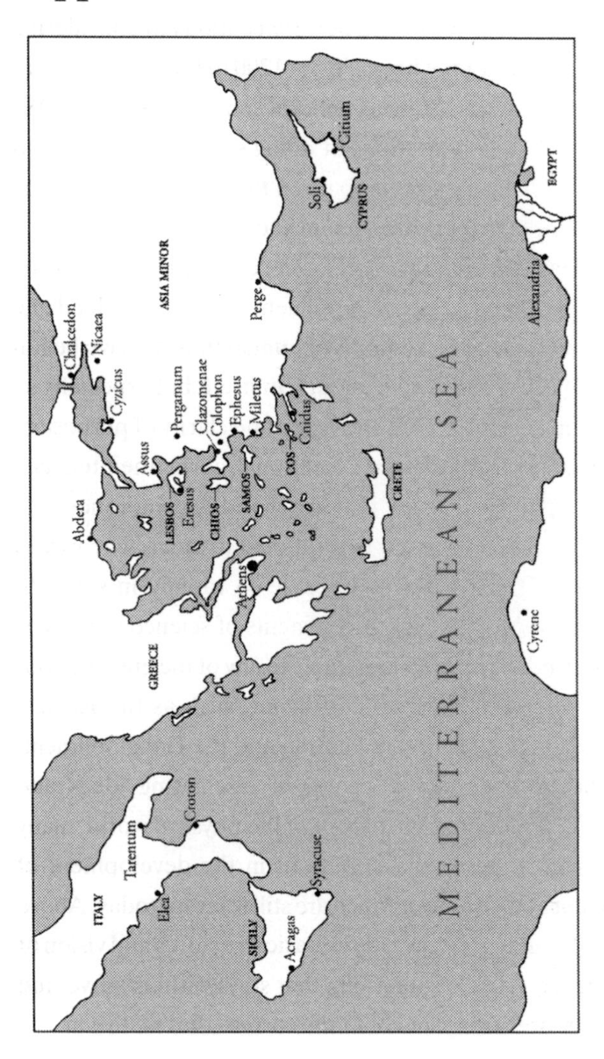

Map of Ancient Greece

Glossary of Terms

Academy: The school of philosophers founded by Plato.

Actual: In Aristotle's theory of explanation, things had a **potential** which they would actualise.

Aether: In Aristotle's cosmology, the fifth element which makes up the celestial realm.

Alchemy (ancient): The art of transforming less valuable or useful things into more valuable or useful things; a wider art than the transmutation of base metals into gold.

Apparent motion: (Apparent) motions of the heavenly bodies which are in fact due to the motion of the earth.

Astrology (ancient): The study of the effect of the heavens on the earth; broader than the modern conception of astrology.

Atoms: Pieces of matter which could not be divided any further (from the Greek *atomos*, uncuttable).

Babylonians: Important culture prior to the Greeks, also known as the **Mesopotamians**, living between the Tigris and the Euphrates rivers (modern Iraq).

***bibbu*:** Babylonian for sheep; used for planets as well, since it appeared that the planets wandered across the night sky.

Black bile: One of the four humours of the **humoural system**.

Celestial realm: In Aristotle's cosmology, the realm from the moon outwards, composed entirely of **aether**.

Centrifocal cosmology: A cosmology in which there is a central point to the cosmos and the natural motion of objects is relative to that point. Contrast **parallel cosmology**.

Concentric sphere astronomy: A view in which the motions of the heavenly bodies are conceived of as combinations of regular circular motion around a common point.

***cosmos* (pl. *cosmoi*):** From the Greek *cosmeo*, to order, with a sense of good order.

***demiourgos*:** Plato's god, who was a craftsman working to geometrical principles when he ordered the cosmos out of chaos.

Ecliptic: Path followed by the sun against the background of the fixed stars, plotted by watching which stars appear at the point on the horizon where the sun sets.

Efficient cause: In Aristotle's theory of explanation, similar to the modern idea of cause.

Enforced motion: In Aristotle's theory of motion, any motion that is not a **natural motion** and requires force.

Epicycle: The basic unit of the astronomy of **Ptolemy**, in which a planet is imagined to be rotating on a sphere which is itself rotating around a central point.

Epicyclic astronomy: The astronomy of **Ptolemy**, which used the epicycle as its basic unit.

Epigenesis: Epigenesis is the view that the embryo develops its different parts from an amorphous beginning, rather than having all of its parts pre-formed. Contrast **pre-formation**.

Equinox: A day of equal night and day.

Final cause: In Aristotle's theory of explanation, teleological explanation.

Formal cause: In Aristotle's theory of explanation, an explanation which cites the form that an object has.

Four causes: The material, final, efficient and formal causes which make up Aristotle's theory of explanation.

Four humours: Blood, black bile, yellow bile and phlegm. When these were in balance the body was healthy; disease was an imbalance of these humours.

Geocentrism: The view that the earth is at the centre of the cosmos.

Geometrical atomism: Plato's atomism, in which the atoms are conceived as being perfect geometrical shapes.

Geometry: Literally land measuring (*ge*, earth, *metreo*, to measure), but developed by the Greeks into a rigorous theoretical system.

Harmony of the heavens: The Pythagorean idea that the heavens moved in such a way as to create a harmonious sound.

Heliocentrism: The view that the sun is at the centre of the cosmos.

Hellenistic period: The period after the deaths of Aristotle (322 BC) and Alexander the Great (323 BC).

***hippopede*:** The shape, like an 8 laid on its side, made by two of the spheres in Eudoxus' theory of planetary motion. From the Greek meaning a horse fetter. When combined with the motion of the two other spheres, retrograde motion could be reproduced.

Holism: Explanation in terms of wholes, rather than constituent parts.

Homocentric sphere astronomy: See **concentric sphere astronomy**.

Humoural system: A view of the human body whereby certain humours (usually blood, black bile, yellow bile and phlegm) were given importance and were critical in health.

Humoural theory of disease: The view that there were certain humours of the body (usually blood, black bile, yellow bile and phlegm) which when they were balanced led to good health.

***idiotes*:** Early Greek term for a medical layman, from which we derive 'idiot'.

Inequality of the seasons: The number of days between solstice and equinox (which defines a season) are not in fact equal, as discovered by Euctemon and Meton.

Irrational number: A number which cannot be expressed as the ratio of two integers, such as $\sqrt{2}$.

***logos* (pl. *logoi*):** Greek for word, account, or ratio.

Love: In Empedocles' cosmology, the principle by which the cosmos is brought together.

Lyceum: The school of philosophers founded by Aristotle.

Material cause: In Aristotle's theory of explanation, an explanation which cites the material that an object is made of.

Mesopotamians: See **Babylonians**.

Natural motion: In Aristotle's theory of motion, the motion that an object will execute when it is unimpeded. The natural motion of earth and water was down, air and fire up, and aether in a circle.

Natural place: In Aristotle's theory of motion, objects had a natural place. The natural place of earth and water was at the centre of the cosmos, air and fire at the edge of the terrestrial realm, aether in the celestial realm.

Nutritive blood: In Galen's physiology, nutritive blood flowed in the veins and distributed nutrition to the body.

Pangenesis: Pangenesis is the idea that characteristics acquired during life are passed on to the offspring, e.g., if giraffes stretch their necks to feed, then baby giraffes are born with longer necks.

Parallax: In astronomy, the angle between apparent positions in the sky at six month intervals, the difference being generated by the fact that the earth is in motion around the sun.

Parallel cosmology: A cosmology in which objects are believed to fall from the top of the cosmos to the bottom, hence the problem of why the earth does not fall. Contrast **centrifocal cosmology**.

Phlegm: One of the **four humours**.

***planetes*:** The Greek for planet, also meaning a wanderer or a vagabond.

Plenum: A view of the world in which there is no empty space at all. Objects move like a fish swimming in water. Contrast atomism, in which there are small pieces of matter moving in empty space.

***pneuma*:** In Stoic cosmology, the active principle, also associated with fire and god.

Potential: In Aristotle's theory of explanation, things had a potential which they would **actualise**.

Precession of the equinoxes: Because the axis of the earth's rotation itself has a slow motion (taking 26,000 years to complete a cycle), the position of the night sky looks very slightly different with successive equinoxes.

Pre-formation: The view that embryos are small pre-formed human beings, and merely grow rather than develop parts which they did not initially have. Contrast **epigenesis**.

Pre-Socratic: A philosopher or scientist who worked prior to Socrates (469–399 BC).

Prime mover: In Aristotle's cosmology, god, who is unmoved but causes the motions of the heavenly bodies.

Prognosis: The art of telling what course a disease will take.

Ptolemaic astronomy: The system of astronomy devised by Ptolemy (building on work by Hipparchus and Apollonius) which uses the epicycle as its basic device.

Pythagoras' theorem: In right-angled triangles, the square of the hypotenuse (the longest side) equals the sum of the squares of the other two sides.

Qualitative cosmology: A view of the world which holds that qualities (hotness, wetness, etc.) are the basic items, and are irreducible. The world is best described in terms of these qualities. Contrast **quantitative**.

Quantitative cosmology: A view of the world in which quantities can best be used to describe it, qualities reducing to matter and motion which can be treated quantitatively. Contrast **qualitative**.

Reductionism: A type of explanation whereby certain entities are said to be no more than matter and motion – so heat is no more than particles in rapid motion.

Retrograde motion: An apparent reversal of the motion of a planet against the background of the fixed stars.

Sacred disease: Epilepsy, thought by many in the ancient world to be due to some sort of possession by the gods.

Solstice: The shortest night/longest day, or vice versa, and also the day on which the setting point of the sun stops moving on the horizon and then returns in the opposite direction.

Strife: In Empedocles' cosmology, the principle responsible for the dissolution of order in the cosmos.

Taxonomy: The science of classifying living organisms.

Teleology: Literally an end-directed explanation, typically the end being some good.

Teleology (Aristotle): Teleology for Aristotle is inherent in nature.

Teleology (Plato): Teleology for Plato is imposed upon nature by the *demiourgos* when he orders the cosmos from chaos.

Terrestrial realm: In Aristotle's cosmology, the earth and the region up to, but not including, the moon; composed of earth, water, air and fire.

TOE: In modern cosmology, a theory of everything.

Unmoved mover: In Aristotle's cosmology, god, who is unmoved but causes the motions of the heavenly bodies.

Vivified blood: In Galen's physiology, blood which carries the 'vivifying' spirit from the lungs, through the arteries to the rest of the body.

Yellow bile: One of the four humours in the humoural system.

Zeno's paradoxes: A set of paradoxes in which motion and change are shown to be impossible.

Zodiac: A band across the sky, either side of the ecliptic, through which the planets move.

Timeline of Ancient Greek Philosopher-scientists

Thales of Miletus (*fl.* 585 BC) Anaximander of Miletus (*fl.* 555 BC) Anaximenes of Miletus (*fl.* 525 BC)	First philosopher-scientists.
Pythagoras of Samos (*fl.* 525 BC)	Geometry, relationship between maths and physics.
Xenophanes of Colophon (*fl.* 520)	Critical theologian-philosopher.
Heraclitus of Ephesus (*fl.* 500 BC)	Philosopher-scientist.
Parmenides of Elea (*fl.* 480 BC) Zeno of Elea (*fl.* 445 BC)	Eleatic philosophers, interested in the question of change.
Anaxagoras of Clazomenae (*c.* 500–428 BC) Empedocles of Acragas (492–432 BC)	Philosopher-scientists; worked on theory of matter and cosmology.
Leucippus of Miletus (*fl.* 435 BC) Democritus of Abdera (*fl.* 410 BC)	The first atomists.

Archytas of Tarentum (*fl.* 385)	Follower of Pythagoras.
Hippocrates of Cos (*c.* 460–370 BC)	Founder of rational medicine.
Euctemon and Meton (Athens, *fl.* 430 BC)	Astronomers, discovered inequality of seasons.
Socrates (469–399 BC) Plato (427–348 BC) Aristotle (384–322 BC)	Great Athenian philosophers.
Theophrastus (371–386 BC)	Follower of Aristotle, worked on life sciences.
Eudoxus of Cnidus (*fl.* 365) Callippus of Cyzicus (*fl.* 330)	Astronomers, improved models of the heavens.
Euclid (*fl.* 300 BC)	Founder of Euclidean geometry.
Epicurus of Athens (*c.* 342–271 BC)	Atomist philosopher.
Zeno of Citium (335–263 BC) Cleanthes of Assus (331–232 BC) Chrysippus of Soli (*c.* 280–207 BC)	Founders of Stoicism.
Erasistratus of Chios (*fl.* 260 BC) Herophilus of Chalcedon (*fl.* 270 BC)	Important work in anatomy and physiology.
Archimedes of Syracuse (287–212 BC)	Mathematics, physics and 'Eureka'!
Eratosthenes of Cyrene (*c.* 276–195 BC)	Estimation of size of the earth.

Apollonius of Perga (262–190 BC) Hipparchus of Nichaea (*fl.* 135 BC)	Improvements in astronomy.
Hero (or Heron) of Alexandria (*fl.* 60 AD)	Technology, first crude steam engine.
Ptolemy of Alexandria (*c.* 100–170 AD)	Greatest astronomer of antiquity.
Galen of Pergamum (*c.* 129–200 AD)	Greatest anatomist and doctor of antiquity.

Further Reading

By far the best general introductions to ancient science are G.E.R. Lloyd's *Early Greek Science to Aristotle* (1970) and *Greek Science after Aristotle* (1973).

A little old, but still a useful introduction, is S. Sambursky, *The Physical World of the Greeks* (1956).

Philosophy and science were very much intertwined in the ancient world. A useful introduction to Greek philosophy is the *Routledge History of Philosophy*, vols 1 (1997) and 2 (1999), and a more extended introduction can be found in W.G.C. Guthrie's *History of Greek Philosophy* in six volumes (1962–81).

The Cambridge Companions series is a useful way of finding out current thinking on various philosophers. While they concentrate on philosophy, there is always some useful material on ancient science as well. There are Cambridge Companions to *Early Greek Philosophy*, ed. D. Sedley (1999); *Plato*, ed. R. Kraut (1992); *Aristotle*, ed. J. Barnes (1995); and *Hellenistic Philosophy* (forthcoming).

The standard work on the pre-Socratics, useful for both their science and philosophy, and giving a good amount of the original material, is G.S. Kirk, J.E. Raven and M. Schofield's *The Presocratic Philosophers* (1983).

Somewhat more difficult and controversial, but very entertaining, is J. Barnes' *The Presocratic Philosophers* (1979).

On Plato, A.D. Gregory's *Plato's Philosophy of Science* (2000) is the latest work, or there is also G.L. Vlastos' *Plato's Universe* (1975), which has an interesting introductory chapter on early Greek cosmology.

On Aristotle, the best introduction is G.E.R. Lloyd, *Aristotle: The Growth and Structure of his Thought* (1968), especially chapters 6 and 7.

More advanced on Aristotle, and also containing a great deal of material on issues in science and philosophy in the ancient world, are R.R.K. Sorabji's *Necessity, Cause and Blame* (1980), *Time, Creation and the Continuum* (1983) and *Matter, Space and Motion* (1988).

For thinkers coming after Aristotle, R.W. Sharples' *Stoics, Epicureans and Sceptics* (1996) is a good introduction to Hellenistic philosophy and science, while Long and Sedley, *The Hellenistic Philosophers*, 2 vols (1987) gives a great amount of original material along with a commentary.

On specific subjects in ancient science, for cosmology I would recommend D.J. Furley, *The Greek Cosmologists* (1987) and M.R. Wright, *Cosmology in Antiquity* (1995).

On astronomy, J. Evans, *The Theory and Practice of Ancient Astronomy* (1998) is the latest work, and while T.L. Heath, *Aristarchus of Samos* (1913) is somewhat old, it is still a good introduction.

Somewhat more technical are D.R. Dicks, *Early Greek Astronomy to Aristotle* (1970) and G.J. Toomer, *Ptolemy's Almagest* (1984).

On mathematics, the standard work is T.L. Heath, *Greek Mathematics* (1931).

On the life sciences, R. French, *Ancient Natural History* (1994) is the latest work.

On medicine, J. Longrigg, *Greek Rational Medicine* (1993) is very good, as is his sourcebook on Greek medicine, *Greek Medicine from the Heroic to the Hellenistic Age* (1998). On the Hippocratics, see G.E.R. Lloyd (ed.), *Hippocratic Writings* (1987).

For engineering, see J.G. Landes, *Engineering in the Ancient World* (1978) or D. Hill, *A History of Engineering in Classical and Medieval Times* (1984).